D1010707

STREAM

Tiny Plankton, Giant Bluefin, and the Amazing Story of the Powerful River in the Atlantic

Stan Ulanski

The University of North Carolina Press | Chapel Hill

Designed by April Leidig-Higgins
Set in Garamond PremierPro by Copperline Book Services
Manufactured in the United States of America

The paper in this book meets the guidelines for permanence and durability of the Committee on Production Guidelines for Book Longevity of the Council on Library Resources.

The University of North Carolina Press has been a member of the Green Press Initiative since 2003.

Library of Congress Cataloging-in-Publication Data
Ulanski, Stan L., 1946–
The Gulf Stream: tiny plankton, giant bluefin, and the amazing story of the powerful river in the Atlantic / Stan Ulanski.
p. cm.
Includes bibliographical references and index.
ISBN 978-0-8078-3217-2 (cloth: alk. paper)
1. Gulf Stream. I. Title.
GC269.U43 2008
551.46'2131 — dc22 2008004746

A Caravan book. For more information, visit www.caravanbooks.org.

12 11 10 09 08 5 4 3 2 1

University of North Carolina Press books may be purchased at a discount for educational, business, or sales promotional use. For information, please visit www.uncpress.unc.edu or write to UNC Press, attention: Sales Department, 116 South Boundary Street, Chapel Hill, NC 27514-3808.

CONTENTS

ILLUSTRATIONS

PREFACE

I WINCED SLIGHTLY when my fishing companion, Matt, queried me about the effects of the Gulf Stream on the game fish that we were seeking on a charter out of a North Carolina marina. As a professor, I have come to expect, though reluctantly accept, almost a complete unawareness of the Gulf Stream from students in my introductory oceanography class. Even Matt, an experienced angler, had woefully little knowledge about the mighty, warm, and deep current, flowing like a river just off the Atlantic coast. Matt listened eagerly to my explanation of how water temperature gradients of the Gulf Stream concentrate game fish prey, and how this powerful current affects all life forms, from the tiniest bioluminescent plankton to the largest bluefin tuna.

In retrospect, there was a time when my own knowledge of the Gulf Stream was at best scanty, but that started to change more than three decades ago, when I took my first research cruise. Like the mariners of a bygone period, I sailed into the realm of the Gulf Stream, experiencing firsthand this relentless flow of the Atlantic — where cobalt-blue water rushed past our ship, and where tuna and marlin roamed in search of an easy meal.

But my early sea adventure would not start off smoothly; I was seasick most of the way out. I soon learned that queasiness was no excuse for avoiding the routine task of taking measurements at sea. Called on deck by the senior scientist one evening, I was startled to see the bow of the ship aglow with a pale light. Little did I know at that time that large accumulations of microscopic plants, called dinoflagellates, in the Gulf Stream produce "phosphorescent seas," where the waters glow an electric blue. But my enjoyment of the light show was short-lived; the ship growled to a stop. A heavily rolling ship, a heaving stomach, and swinging instruments don't make for a pleasant sea experience. But persevering, I soon came to realize that my measurements yielded one of the defining characteristics of the Gulf Stream: its relatively high temperature, above 80°F (27°C). As I retired to my bunk after my shift, I reflected on the sea adventures of none other than the ever-curious Benjamin Franklin, statesman, diplomat — and scientist. On at least three cruises to the Gulf Stream, he meticulously recorded the temperature of this warm river and contributed to charting its path.

But over the years, I have come to view the Gulf Stream as more than a discrete entity to be probed and measured and rather as a vibrant and dynamic force of nature. The Gulf Stream, the nexus of this story, is an ocean current that links together many strands from both the scientific and cultural worlds. This book explores the science of the Gulf Stream, examines its life-forms, from its drifting organisms to its nomadic riders, and investigates the North American experience of discovery, colonization, and development that depended on it. Hopefully, understanding these complex connections will give readers a greater appreciation of this unique natural phenomenon about which surprisingly few people, including residents of and visitors to the Atlantic coast, are even aware.

The American scientist Matthew Fontaine Maury, the "father of modern oceanography," was most assuredly cognizant of the Gulf Stream. In 1855, Maury probably came closest to getting at its essence when he wrote, "There is a river in the ocean." From the shores of the Atlantic, no sign of this river of water can be discerned. But just over the horizon, a strong flow signals its presence to even the most casual observer. This oceanic river is like no river on land; its size, range, and power dwarf even the mightiest continental river. And although its "banks" are fluid ocean, not soil and rock, it is clearly visible from space. From its tropical origins, the Gulf Stream moves water poleward at a rate hundreds of times the combined flows of the Amazon and Mississippi Rivers. This river, an ocean current, is part of something larger than itself — a great, clockwise-flowing gyre, the main water circulation system of the North Atlantic Ocean.

Water in the Gulf Stream can move for surprisingly long distances within well-defined boundaries, marked by distinct changes in current speed, temperature, and water color. Hugging the eastern seaboard, it flows almost arrow-straight from southern Florida to Cape Hatteras, North Carolina. But further downstream, it traverses the North Atlantic in a meandering, serpentine path. As if alive, it twists and turns in great sweeping loops, some of which pinch off from the main flow — much as oxbow lakes are formed by meanders in a continental river. The results are large, swirling vortices, or rings. Like spinning tops, these rings rotate either clockwise or counterclockwise around a core of seawater that is distinct from the surrounding water. During the course of a year, the Gulf Stream may shed a dozen or more rings to either side of its edges, constantly remaking itself.

What powers the Atlantic's gyre? For centuries, the answer proved elusive, though not for lack of trying — all manner of ideas and speculations flowed from many sectors of society. The philosopher Socrates proclaimed that deep channels in the Earth funneled an unending stream of water around the globe. Even Maury argued, incorrectly, that surface currents like the Gulf Stream result from density differences within seawater. He staunchly believed that global winds were too weak to have any influence on ocean currents. However, even a cursory look through modern earth science textbooks reveals that scientists now recognize that ocean surface circulations do indeed reflect global wind patterns, though not in a straightforward manner. The recipe for a gyre on the scale of an ocean includes the wind drag across the ocean surface, pressure gradients in the ocean, and the influence of the rotating Earth. Add these together, inject some physical and mathematical insight into the coupling between the atmosphere and the ocean, and what emerges is a realistic picture of ocean circulation.

All kinds of riders accompany the Gulf Stream on its northward journey through the western Atlantic. Some of these passengers are passive organisms, literally going with the flow, like the floating sargassum weed. This pelagic, brown algae, often growing in long lines along the Gulf Stream's boundaries, is a critical source of food, shelter, and substrate for a multitude of other marine organisms. At the other end of the spectrum of Gulf Stream travelers is the giant bluefin tuna, which spawns in the Gulf of Mexico and undertakes prodigious feeding migrations of more than two thousand miles. Revered by Mediterranean societies dating back to antiquity, bluefin tuna are marvelous specimens of natural engineering — streamlined bodies, heat-conserving circulatory systems, and sickle-shaped tail fins — who have all the biological tools needed to undertake their long migrations.

Human ingenuity meant that ocean currents also became sea highways for fifteenth- and sixteenth-century explorers. Christopher Columbus found the way to the New World, but Ponce de León found the way back: he discovered the Gulf Stream in 1513, and from then on it was the way home for Spanish galleons laden with the treasures of the Caribbean and Mexico. But the current extracted a steep toll: shipwrecks were common where the Gulf Stream brushes against the reef-strewn Florida Keys. The loss of galleons became the gain of the indigenous Indians of Florida, who were the first to salvage cargo from the wrecked ships. The Spaniards would experience other sorts of trouble, with the burgeoning of piracy in all of its many forms. Bands of sea bandits, known as buccaneers and privateers, roamed the Caribbean and the Gulf Stream in search of ships to plunder and loot. During the golden age of piracy, thousands of men sailed under the "Jolly Roger" banner, and none was more famous than Bartholomew (Black Bart) Roberts, who captured more than four hundred ships in only three years. Today, a new breed of adventurers seek the treasures of the Gulf Stream. Their jewels are the game fish — dolphin, wahoo, and marlin — that cruise its waters in search of an easy meal. From the golden yellows and deep greens of a leaping dolphin to the iridescent blues of a hard-fighting marlin, these colors are as mesmerizing to the big-game angler as the glitter of gold from a Spanish galleon was to a pirate.

This book will show how, just as continental river travel was vital to the economic growth of Renaissance Europe, the Gulf Stream played a major role in the colonization of America. Spanish and Portuguese expeditions in the fifteenth and sixteenth centuries discovered the most favorable winds and currents for reaching the New World: one would sail south to the Canaries, turn right at these islands, ride the trade winds and the lower limb of the gyre across the Atlantic to the land of the Indians, make another right, and sail the Gulf Stream northward to America. Though this route to the New World was considerably longer than the more direct westward course from European ports, it was actually faster.

The sea paths to the Americas blazed by early European explorers would forever transform the New World, allowing for unprecedented migration and exploitation. The Gulf Stream made possible an enormously lucrative trade in spices, sugar, and rum during the colonial era, and it also, horribly, floated ships filled with African slaves, whose forced labor is now acknowledged as having been essential to the building and prosperity of the

American colonies. So important was the Gulf Stream to all of this trade that early maps depicting its course were closely guarded secrets. While plenty of mysteries remain about the Gulf Stream, much more of its story is now known. It is a story that connects the complex network of physics, biology, and human interaction that characterizes our world.

ACKNOWLEDGMENTS

NO ONE REALLY GOES IT ALONE, so I'm indebted to my colleagues, Michael Garstang, Frank Gerome, Eric Pyle, and Kristen St. John, who graciously offered their advice and support as I struggled to define and structure this book.

My thanks to Lacy Rainwater; she produced many of the figures that became an integral part of illuminating the Gulf Stream and its inhabitants.

Institutional support came from the College of Science and Mathematics at James Madison University, which provided me with an educational leave to pursue this fascinating topic.

Finally, sincere thanks to Elaine Maisner, Senior Editor at the University of North Carolina Press, whose confidence in the project remained strong throughout the years.

PART 1 COMING FULL CIRCLE

FLOW IN THE ATLANTIC

1 SWIRLS AND CONVEYORS

LOOK AT A MAP OR A GLOBE of the Earth; either will do. Both depict in great detail the geography of our planet: continents visibly stand out; seas and oceans abut these landmasses. Maps and globes are snapshots of the ordered arrangement of land and water; they are a cartographer's still life. But they yield little information about the earth's dynamic nature.

Two relatively thin but interrelated shells cover the surface of the Earth: the atmosphere, essentially the air, which supports the higher forms of life on this planet, and the hydrosphere, the Earth's water. Each has different physical properties that account for their unique natures but also their propensity to interact. Though the Greek myth tells us that the Titan Atlas was condemned to hold up the sky for eternity, the sky is not falling; air and water are perpetually separated into distinct layers. Both of these realms are fluids that are capable of flowing; neither is static. Clouds scudding above the earth's surface indicate, even to the casual observer, the movement of air. The world's oceans, the major component of the hydrosphere, have waves and tides ceaselessly moving across their surfaces. One type of motion intrinsic to both the atmosphere and the ocean is the horizontal movement of these fluids: winds in the atmosphere and currents in the ocean. If winds are fast, then in comparison currents are slow. The much denser seawater is just plain sluggish.

Unlike the atmosphere, the oceans cover only 71 percent of the planet, with their currents strongly confined by lateral boundaries imposed by the geometry of the continents. While winds like the jet stream girdle the globe, large swirls of water are found within the oceans of the northern and southern hemispheres. Water flows around the ocean basins in closed circulations, known as gyres. These huge, circular currents, enclosing over a million square miles of ocean, are the result of global wind patterns and are common to the Atlantic, Pacific, and Indian Oceans. When inspected from above, the circulation is clockwise in the northern hemisphere and counterclockwise in the southern hemisphere. Viewing the flow of surface water from a geometric perspective is deceptively simple because it represents ocean conditions averaged over a long period of time. While true in the mean, the actual current flow at a particular location and time may be quite different than the average conditions. In particular, the closer to land, the greater is the deviation from the gross gyre pattern. The shape of the coastline and changes in bottom topography, to name just two factors, conspire to distort the oceanic ovals.

Our present understanding of these oceanic gyres has evolved over the decades from intensive study of these physical systems. While the proverbial message in the bottle may have been the sole means of a shipwrecked survivor communicating with the outside world, oceanographers do indeed use floating objects to study ocean currents. The Lagrangian method (after the Italian mathematician Joseph Lagrange, 1736–1813, who developed its underlying theory) involves the release of floats that faithfully follow a moving parcel of water. In its simplest manifestation, oceanographers release small drift bottles or drift packets. Each bottle/packet contains a card asking the finder to note the date and location where it was found and to return it.

Initiated in the year 1802 aboard the English ship HMS *Rainbow*, these first bottle experiments, designed to study the current structure of the North Atlantic, continue basically unchanged two centuries later. A number of years ago, the U.S. Coast and Geodetic Survey used drift bottles to study the current pattern in the western Atlantic. A bottle released near Caracas, Venezuela, reached the Florida Keys four months later, traveling at an average speed of sixteen miles per day.

Even Hollywood has gotten into the drift bottle act, to spin a love story. In a 1999 film, appropriately titled *Message in a Bottle*, a young woman walking along a deserted stretch of the Maine coastline finds a passionate letter enclosed in a bottle. She is so moved by the letter's poetry that she seeks out the author, and her quest leads her to the Outer Banks of North Carolina, to a sailboat builder. While I won't dwell on what transpires in

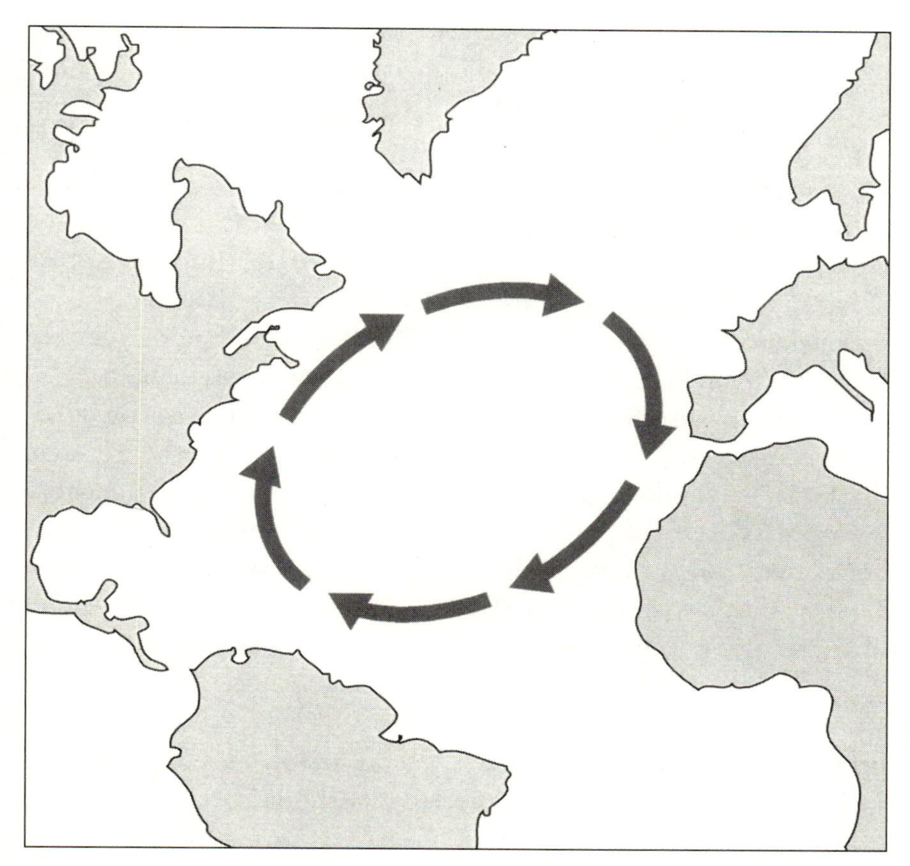

Simplified view of an ocean gyre

the movie, the question relevant to our discussion is, could this bottle have drifted hundreds of miles northward from its origin? Probably, but the love-starved North Carolinian would have had to set the bottle adrift in one of branches of the clockwise gyre in the North Atlantic. (We'll shortly see the specific parts of this gyre.)

Serendipitous studies have, at times, yielded valuable information about the nature of current flow. In January 1992, a merchant ship encountering storm conditions near the International Date Line in the North Pacific lost twelve containers overboard due to the heavy seas. Part of this cargo was 29,000 floatable, plastic bathtub toys: turtles, frogs, beavers, and, yes, ducks. Some of these toys began coming ashore in southeast Alaska ten months later. This unfortunate accident became a scientific gold mine for Curtis Ebbesmeyer and James Ingraham, two University of Washington oceanographers. Using a computer model, they were able to develop a number of possible drift trajectories of the toys as a function of the current's speed and direction. The model indicated the North Pacific gyre would disperse the toys over the next two years throughout the Pacific Ocean, some washing up in Hawaii and others in the frigid Arctic. Since 1992, these providential discoveries of the nature of ocean currents have continued as oceanographers tracked other floating objects, including 34,000 hockey gloves, 5 million Lego pieces, and at least 3,000 computer modules.

The Lagrangian technique most widely used is the deployment of sound-transmitting floats, which can be tracked acoustically from on board a ship. The density of the float is adjusted so it is neutrally buoyant, meaning that it floats at a designated depth. The plot of the drift paths of the floats yields flow patterns: loops, ovals, and figure eights. We are learning—whether by studying the drift of random junk or analyzing the paths of sophisticated instruments—that surface ocean currents can be quite complicated.

The Eulerian method (after the Swiss mathematician Leonhart Euler, 1707–83) involves measuring the current with a meter attached to a cable between a buoy at the surface and an anchor on the ocean floor. After being set in place, the meter is left for a predetermined period—days, weeks, or months—depending on the research objectives. Technicians may suspend a series of current meters at different depths on the cable to obtain a vertical profile of the current's speed and direction.

Did I mention that the ocean is dynamic, marked by vigor and energy? I should probably also include that, at times, the ocean can be quite fickle, thwarting our best efforts to decipher her secrets. While on a research project in the Caribbean, I experienced firsthand her capricious nature.

Personnel from Florida State University and the National Oceanic and Atmospheric Administration had meticulously positioned an instrumented buoy off the eastern coast of the island of Barbados, with the intent of collecting weather and sea data, including current flow. Once the buoy was in place, they would periodically go out to service it and retrieve the data. One day, after a night of congratulatory revelry (the project was proving to be quite successful), I tagged along with them, bouncing along in a rubber Zodiac. (My role in the project was essentially shore-based, but during that festive night, I became an honorary Barbadian sea dog.) While each cresting wave allowed an unobstructed view of the horizon, I couldn't see the buoy. We were right on the mark, but the buoy was nowhere in sight. We checked and double-checked our coordinates, but we could only come to the sickening conclusion that it was gone. Had a multi-ton buoy simply vanished without a trace? The sea offered no answers.

I thought of the Vikings who, with more than four dozen gods and goddesses to choose from, could easily find a deity to blame their seafaring misfortunes on. One likely candidate who particularly appealed to me was Aegir, the god of the sea. He was the personification of the sea, be it good or evil. When angered, he would hurl down storms upon the offending mariners. Norse storytellers said that when a ship went into "Aegir's wide jaw," all aboard were taken to his hall at the bottom of the sea. I became convinced Aegir was responsible for the loss of our buoy since prudent Vikings attempted to appease him by offering up human sacrifices — a glaring omission on our part. Though we never determined the fate of the buoy, I took some consolation in the writings of J. G. Kohl, a nineteenth-century German cartographer: "It has been often sayd with truth, that these oceanic currents are the most deceitful things in the world and that it is extremely difficult to become aware of them and to take them into account."

In spite of setbacks — like the one we experienced — analysis of the data from long-term current measurements has shown that each gyre is made up of four, more or less separate, prevailing currents. This division is somewhat arbitrary because there are no distinct geographical boundaries separating these currents, no well-defined beginnings or ends. The distinction relies on the physical dimensions and characteristics of each of these currents. One would be hard-pressed to define the exact source of a current; its "headwaters" are not visible on any map. The waters of a gyre are connected, linked together in a great oval; so, for example, the Gulf Stream, the Canary Current, the North Equatorial Current, and the North Atlantic Current are the links of the subtropical North Atlantic gyre. These interlocking currents surround the borderless Sargasso Sea and isolate this body of water from the coastal waters near the continents. With an area of more than a million square miles, the size of Australia, the relatively stagnant Sargasso Sea stands in sharp contrast to its neighboring currents.

A tour of this gyre might begin with the North Equatorial Current,

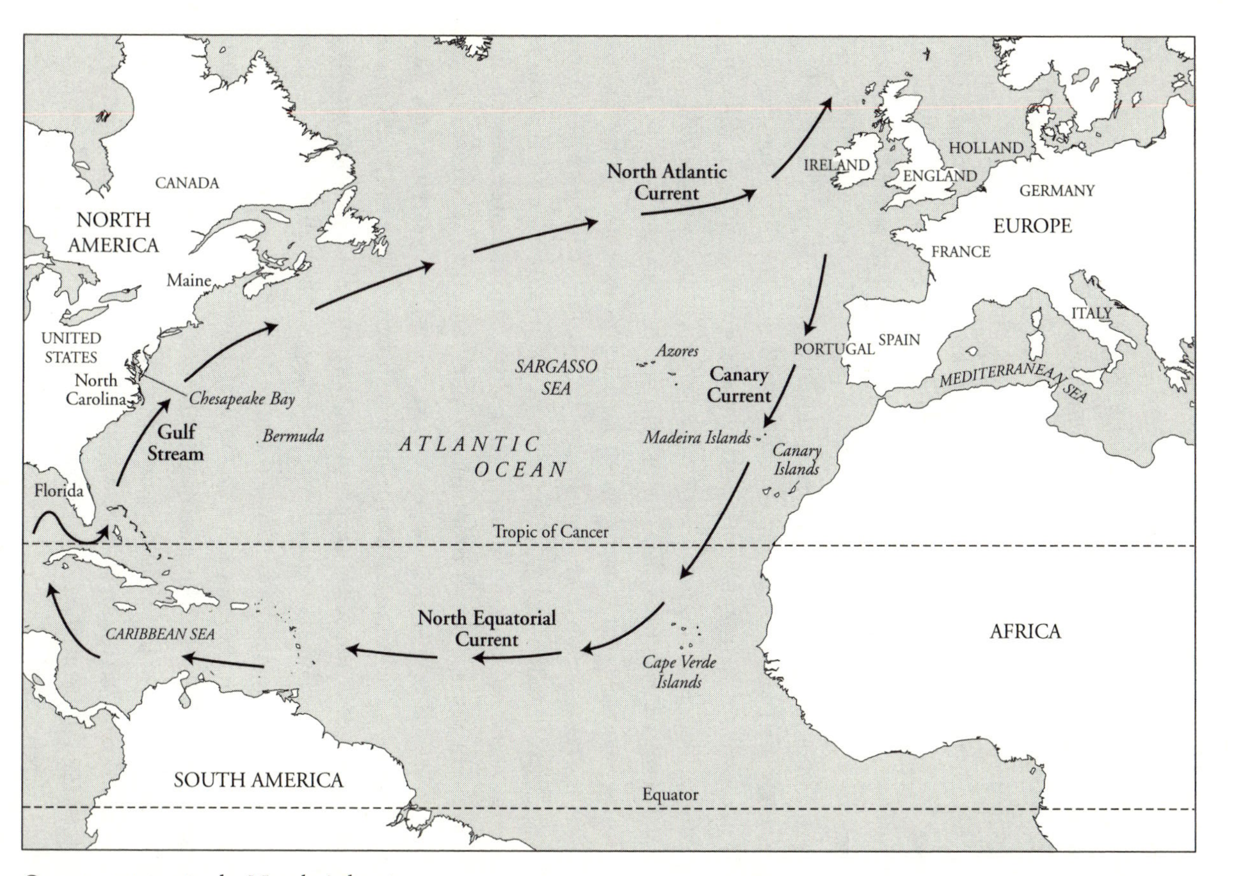

Ocean currents in the North Atlantic

which is found from about 7° to 20° north latitude. Fortified by the Atlantic trade wind belt (0° to 30° north), the North Equatorial Current is a broad (eight-hundred-mile-wide), westward-flowing current that forms the southern limb of the gyre. The current originates off the northwestern coast of Africa, where waters flowing southward from the northeast Atlantic feed into it. As this current travels across the vast expanse of the tropical ocean, waters originating south of the equator join it and contribute to the overall transport of tropical water. With an average speed of less than half a knot (half a nautical mile per hour), the North Equatorial Current flows slowly across three thousand miles of open ocean. (A rubber duck, floating in this current and unaided by winds, would cross this vast expanse of the Atlantic in 250 days.)

When the North Equatorial Current approaches the Americas' continental shelf — a shallow, near-horizontal seafloor extending from the coast to a depth of about four hundred feet — interaction with the bottom topography produces a complicated flow regime. The overall flow is to the northwest, where it splits into two branches: one enters the Caribbean Sea, and the other flows north and east of the West Indies.

In the Caribbean, the closely spaced chain of islands, reefs, and sills (submerged ridges) of the Lesser Antilles acts as a porous barrier for the inflow of Atlantic water into this basin by impeding the movement of deep water. Hydrographic surveys indicate that water flows into the Caribbean Sea mostly through the Grenada, St. Vincent, and St. Lucia passages. The water then continues westward as the Caribbean Current, the main surface current in the Caribbean basin. Near Honduras on the Mexican coast, the current abruptly turns northward and surges through the Yucatan Channel between Mexico and Cuba to enter the Gulf of Mexico.

As far back as 1890, John Pillsbury, a lieutenant in the U.S. Navy, took direct current measurements in the narrow Yucatan Channel and reported a strong flow of more than three knots. The flow of water that intrudes into the Gulf of Mexico is known as the Loop Current, a robust, clockwise circulation that extends northward into the gulf. Occasionally, the Loop Current will reach as far north as Florida's continental shelf before exiting through the Straits of Florida.

The second branch, the Antilles Current, was first named by the Ger-

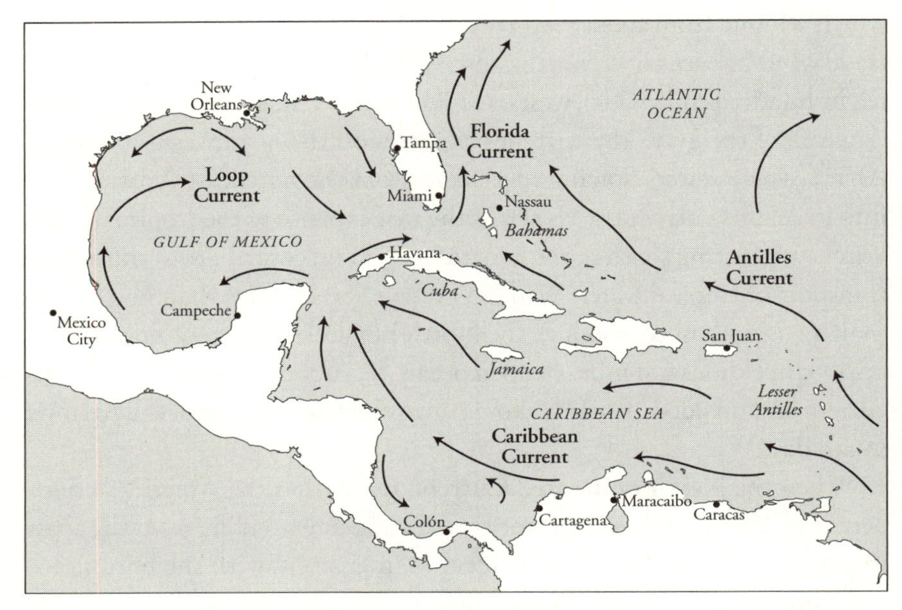

Currents in the Caribbean and the Gulf of Mexico

man oceanographer Otto Krummel in 1876. Subsequent observations in the 1930s by Columbus Iselin from Woods Hole Oceanographic Institution called into question the existence of this current. The discontinuous nature of the current, as well as its weak flow, led to continual speculation about it throughout the 1970s. But by the 1990s, detailed current measurements conclusively confirmed its existence.

The two branches rejoin near southern Florida to form the Florida Current, which many view as the "official beginning" of the Gulf Stream. The current heads virtually north from the passage between West Palm Beach, Florida, and Grand Bahama Island to the Mid-Atlantic Bight (between Cape Canaveral and Cape Hatteras). About ninety miles southeast of Charleston, South Carolina, the Gulf Stream flows directly over the Charleston Bump, a series of underwater scarps, rocky ridges, overhangs, and caves. Though the Charleston Bump is only a small topographical feature on the ocean floor, it exerts a strong influence on the Gulf Stream, disrupting it and deflecting the current offshore. As the western limb of the North Atlantic gyre, the Gulf Stream flows some seven hundred miles from Key West to Cape Hatteras, generally hugging the eastern coastline.

Upon reaching Cape Hatteras, the Gulf Stream enters deeper water, where it flows northeast to the colder climes. As it brushes past the Grand Banks, south of Newfoundland, it turns abruptly eastward under the influence of the prevailing westerly winds. At about 50° west, the Gulf Stream metamorphoses into the North Atlantic Current (North Atlantic Drift); at least, that is the opinion of many who view this current, based upon changes in water properties, as the end of the Gulf Stream. Regardless of the exact terminus of the Gulf Stream, the northern limb of the gyre flows more than two thousand miles toward Europe. As the North Atlantic Current approaches Ireland, it splits into two branches: one branch feeds the Norwegian Current that flows along the Scandinavian coast, and the second branch, the Canary Current, travels south along the west coast of Africa from Morocco to Senegal. The Canary Current ultimately unites with the North Equatorial Current, closing the loop of this great clockwise-flowing gyre.

As we will see in later chapters, the North Atlantic gyre would become the sea highway for Europeans traveling to far-off destinations. From Columbus, sailing the North Equatorial Current to the New World, to Ponce de León, "discoverer" of the Gulf Stream, to the Portuguese explorer Gil Eanes, riding the Canary Current to remote reaches of Africa, men who sailed the sea would become intimately familiar with the flow of water in the Atlantic. Unfortunately, these early adventurers produced little permanent evidence, such as maps and charts, depicting their sea routes. Dutch cartographers, American merchants, and Spanish treasure fleet captains would fill the void, producing rudimentary maps of the Atlantic circulation.

Modern-day cartographers take pride in producing maps that show the correct position of continents, oceans, and gyres. But the Earth is not as static as it is represented in even the most handsomely bound atlas. About 200 million years ago, the geography of our planet did indeed look markedly different than it does today. Landmasses were locked together in the supercontinent scientists call "Pangaea," which was surrounded by a vast ocean they call "Panthalassa." The Atlantic Ocean and its circulation did not yet exist and would not exist for millions of years to come.

The breakup of Pangaea and the subsequent drift of these massive fragments produced large changes in the geography of the planet. As global seaways opened or closed, allowing water to flow where it once was blocked by

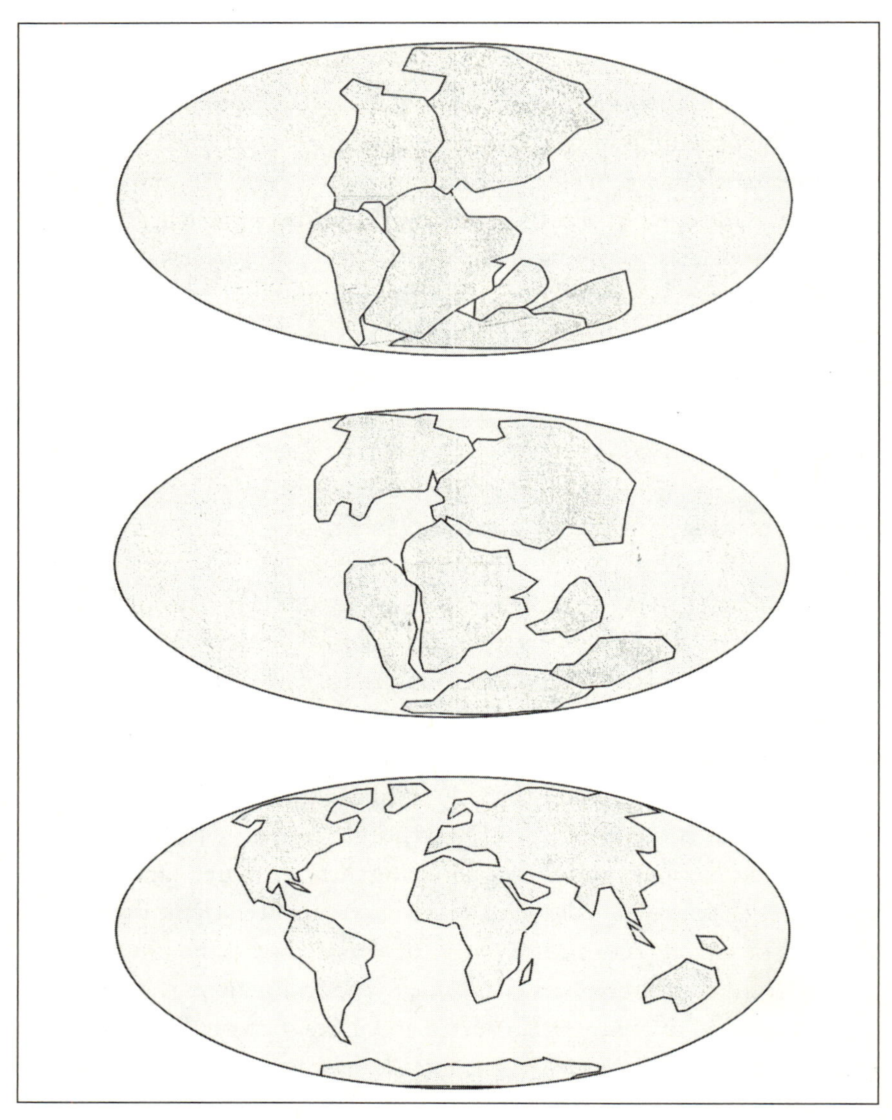

Opening of the Atlantic Ocean, 200 million years ago to the present

land, the ocean circulation began to take form. Geological, oceanographic, and other geophysical evidence allow us to reconstruct this evolving mosaic that took place over millions of years.

From the first 75 million years of the fragmentation of Pangaea, very little is known about the global ocean circulation. Pieces of the puzzle only point to a broad, sluggish flow throughout Panthalassa. The Atlantic Ocean was just starting to open.

Some 125 million years ago, when dinosaurs roamed the earth, the North Atlantic was still isolated from the Arctic, and blocks of North America remained joined with the massive Eurasian landmass. With the further widening of the seaway along the equator, a globe-encircling equatorial current system may have developed. Unrestricted by landmasses, unlike the present-day North Equatorial Current, this current was probably quite impressive, flowing in a broad band tens of thousands of miles around the earth. Since there were no barriers to deflect the equatorial current, very little water was probably diverted to the northern and southern hemispheres.

The next stage, 60 million years ago, marked the opening of all oceans, though they would not yet have attained their present-day dimensions. With the continual drift of the continents, the role of these landmasses in deflecting currents took on a greater significance. In particular, a branch of the aforementioned equatorial water now flowed south into the embryonic Indian Ocean to begin its long journey around Africa.

Thirty million years ago, the oceans began to take on an appearance and shape that would be recognizable to present-day voyagers. Still not completely filled out as we know it today, the Atlantic Ocean now stretched unimpeded all the way from the tropics to the poles. The North Atlantic gyre became a prominent and permanent feature from this period forward. The newly formed Gulf Stream now received a considerable portion of its water from the Atlantic Equatorial Current, a residual segment of the globe-girdling equatorial current system of 125 million years past. From this time on, life in the Gulf Stream, from the microscopic plankton to the giant bluefin tuna, was able to ride this great current throughout the western Atlantic.

While the subtropical gyre has taken center stage in the Atlantic for millions of years, the Gulf Stream is its star. With a flow that dwarfs any continental river, the Gulf Stream is a mighty oceanic river, powerful enough to be readily seen from space. This current-as-river analogy dates back to 1855 when Matthew Fontaine Maury, in *The Physical Geography of the Sea and Its Meteorology*, aptly described the Gulf Stream: "There is a river in the ocean. In the severest droughts it never fails and in the mightiest floods it never overflows; its banks and bottom are of cold water, while its current is of warm, the Gulf of Mexico is its fountain, and its mouth is the Arctic Sea." Present-day oceanographers still find this analogy strikingly apt: Water in the Gulf Stream can move a surprisingly long distance, a hundred miles in a day, within well-defined boundaries characterized by dramatic changes in physical and chemical properties. These boundaries, which mark the width of the Gulf Stream, are relatively narrow, stretching no more than sixty miles in an east-west direction.

Within these watery boundaries, like the banks of a continental river, the flow is fast and intense, with speeds of more than four knots — making the Gulf Stream one of the swiftest currents in the world's oceans. Studies have shown that the Gulf Stream, as it flows past Cape Hatteras, maintains its relatively high speed to a depth of approximately a thousand feet, and water flow, albeit significantly reduced, still exists more than six thousand feet below the surface. This narrow, fast, and deep current is one of five western boundary currents found along the western edge of every ocean basin. In contrast, eastern boundary currents, like the Canary Current, are wide, slow, and shallow. The jetlike structure of the western versus the eastern boundary currents is a dynamic response to the wind, pressure fields of the ocean, friction between the current and the bordering landmass, and the deflecting forces produced by the rotation of the Earth.

The Gulf Stream moves a prodigious amount of water poleward through the western ocean basin. A visitor to the Cape Hatteras National Seashore might be surprised to find out that this current, located less than forty miles from the coast, transports more than 2 billion cubic feet of water every second. That is approximately *five hundred times* the transport capacity of the Amazon River, which supplies one-sixth of all the fresh water discharged to the world's seas. This large transport off Cape Hatteras is more than twice that off the coast of Florida. Scientific evidence suggests that this difference is due to high deep-water velocities near Hatteras, coupled with a large volume of water entrained from the Sargasso Sea.

A snapshot view of the Gulf Stream can be misleading to an observer, giving the impression that the current is stagnant over space and time. Nothing could be further from the truth; the path of the Stream changes constantly downstream of Cape Hatteras, month to month and certainly year to year. The stream "wobbles," and immense, wavelike oscillations form. As if alive, it snakes across the western Atlantic. The wavy patterns are meanders, analogous to the broad, curved loops of river channels winding their way across floodplains. (The word "meander" derives from Menderes, a river in Turkey that has a very winding course.) Initially, the size of these meanders is not impressive; the fluctuations have a relatively small amplitude of some 30 miles. But further downstream of Cape Hatteras, the meanders, like ocean waves in a storm, grow significantly in amplitude (up to 150 miles) and in wavelength (200 miles). Similar to a wave progressing down a taut string, the meanders propagate down the Gulf Stream at a slower but separate speed from the main current flow.

As the swift-flowing Gulf Stream twists and curves in great sweeping arcs, a meander may bend sharply, forming almost a complete loop. Such a pinched loop may be cut off from the main flow, much as an oxbow lake forms from a meander in a river, resulting in a closed, self-contained vortex or ring. As each ring forms, a column of water is appropriated from one side of the current and transported to the other side, a region of distinctly different properties, both physical (e.g., cold rings and warm rings, discussed in Chapter 2) and biological (planktonic organisms, discussed in Chapter 4). Rings can form either to the north (landward) or south (seaward) of the Gulf Stream. A ring that forms to the north will have a core of water that originated from the south in the Sargasso Sea. Conversely, a ring that is found to the south has its origin in a region called the Slope Water, which extends from Cape Hatteras to the Grand Banks of Newfoundland. These rings have diameters of one to two hundred miles — yielding an area comparable to that of New Hampshire, Vermont, and Rhode Island combined — and may spiral all the way down to the seafloor. Once spun off from the Gulf Stream, these rings may persist on average for one or two years as they slowly drift (two to four miles per day) to the southwest, opposite the Gulf Stream flow. Gradually decaying along this path, the rings enter a region north of the Bahamas and east of Florida where they coalesce with the Gulf Stream and disappear. On average, the Gulf Stream in remaking itself may shed dozens of rings in a single year.

The dynamic nature of the Gulf Stream presents a major challenge to any serious sailor bent on winning an open-ocean, long-distance race. In

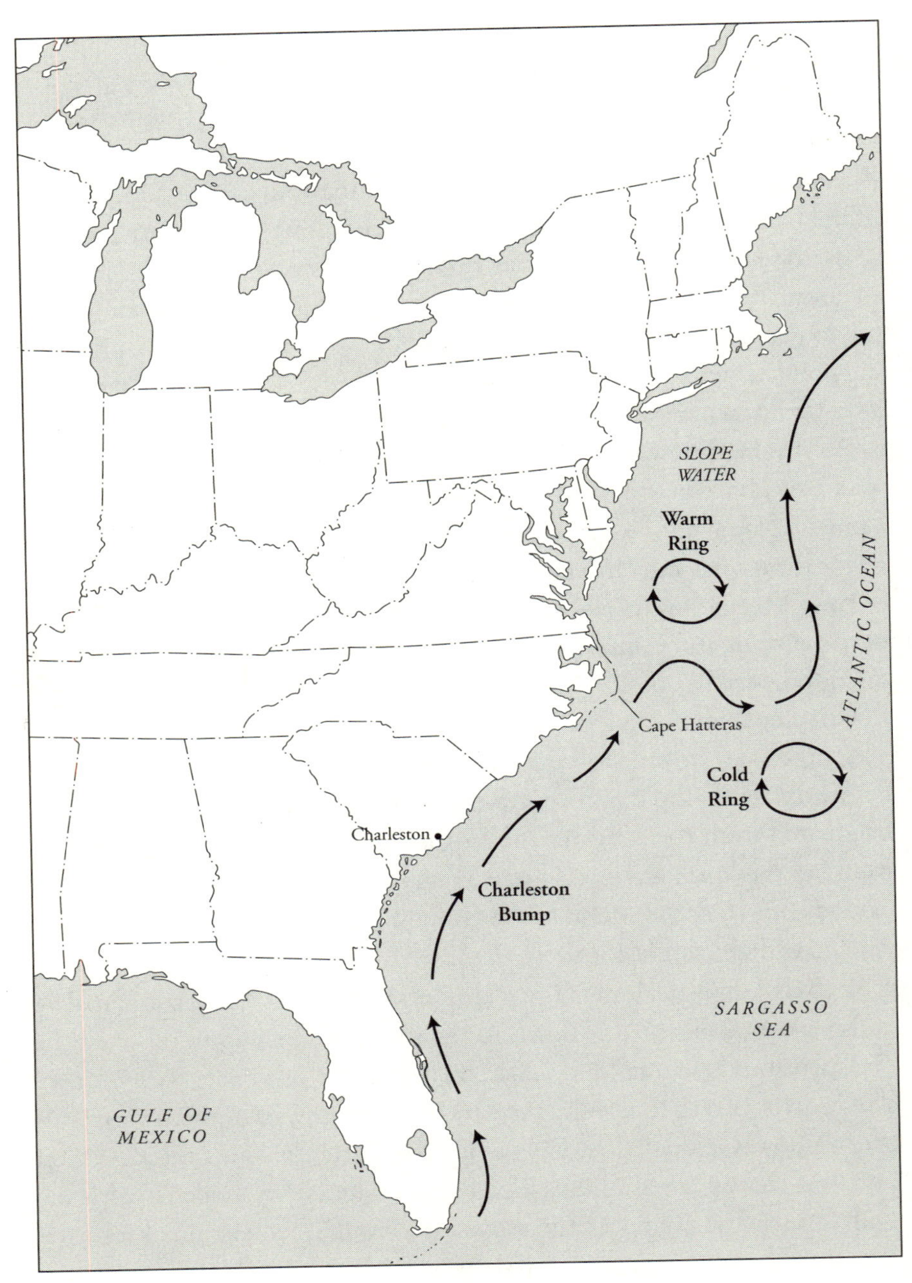

Gulf Stream meanders and rings off the mid-Atlantic coast

particular, the prestigious Newport, Rhode Island, to Bermuda race (635 miles) involves being able to take advantage of the quirks in current speed and direction. Optimal use of the Gulf Stream can mean the difference between finishing first or being left out of the awards ceremony. Tactical considerations include assessing the relationship among current, wind, and waves. A wind that is opposing the stream's direction can cause the sea to build up, leading to dangerously high waves.

Locating, tracking, and navigating around the whirling rings are additional keys to success. Rings found north of the Gulf Stream rotate in a clockwise fashion. A vessel sailing from Newport that finds itself on the eastern side of the ring would have a southward flow of water at its stern — sluicing it toward Bermuda.

The yachtsman unfortunate enough to be on the western side of the ring would have an opposing current at the vessel's bow, markedly negating any headway gained as a result of the wind. In the 1986 race, sailboat navigators recorded rotational speeds of more than seven knots for some of these rings. The intense rotational flow literally pushed one unlucky competitor, who found himself on the wrong side of the ring, backward! Upon crossing the Gulf Stream, sailors have to adjust their tack to contend with the counterclockwise rotating rings, which are found south of the current.

While modern oceanographic techniques allow scientists to locate and track rings, an ancient map of the North Atlantic, drawn more than 450 years ago, may have been the first depiction of these ocean phenomena. The map is the *Carta Marina*, which Olaus Magnus, an exiled Swedish priest living in Italy, published in 1539. Taking more than twelve years to complete, the *Carta Marina* is an ornate and detailed map of the Nordic countries from Iceland to Finland. Though it contains a wealth of information about the towns, lakes, and people of each country, the map devotes a similar richness of information about the ocean. What particularly caught the eye of Tom Rossby from the University of Rhode Island was the way Magnus had drawn a broad band of ocean swirls in only one region of this vast ocean area. This band stretched from Iceland eastward to Norway, passing just north of the Faroe Islands. Was there any significance to these precisely located swirls, or was Magnus simply expressing his intense dislike of a blank canvas, covering available space with other objects such as sea snakes, sinking galleons, and sea monsters?

Carta Marina by Olaus Magnus (courtesy of NOAA)

Rossby thought the location, size, and spacing of the swirls were drawn too deliberately to be mere artistic license. Rossby was quick to realize that they corresponded almost perfectly with the Iceland-Faroes Front, the boundary separating the Gulf Stream from the cold Arctic water. To verify his hunch, Rossby needed observations of this region. Satellite images revealed the existence of ocean eddies — slowly rotating masses of water up to sixty miles in diameter but smaller than Gulf Stream rings — along the Iceland-Faroes Front. At the point where the Gulf Stream clashes with the Arctic water, eddies spin off from this junction. But without the advantage of satellite images or other oceanographic measurements, how did Magnus obtain this information? It is possible the artist gleaned these facts from the traders and mariners who sailed the waters of the North Atlantic. In particular, the mariners of the Hanseatic League, who traded staples between Germany and Iceland, were experienced sailors and would have observed that their slow-moving vessels were pushed off course by more than thirty degrees when crossing the front. They would have dutifully recorded this information as an aid to navigation.

Olaus's depiction of ominous sea creatures attacking the trading ships also deserves some attention. If the artist rendered the swirls accurately, then can we accept the existence of sea monsters roaming the North Atlantic? As a Nordic inhabitant, Magnus would have most likely been familiar with the twelfth-century Norwegian stories about Kraken. The legend of Kraken refers to a creature the size of an island that would attack a ship, reaching as high as the top of a ship's mast with its long arms and ultimately pulling the vessel down to the bottom. Though tales of Kraken persisted even in 1752, when Erik Ludvigsen Pontoppidan wrote about the Kraken in his book *The Natural History of Norway*, the Kraken of legend was probably the giant squid. Though obviously smaller than a floating island, giant squid can reach lengths of more than fifty feet and weights approaching a ton. The giant squid is generally reclusive, a denizen of the deep that resides in the ocean depths ranging from six hundred to three thousand feet. The poet Alfred, Lord Tennyson reinforced this theme of isolation when he wrote:

> Below the thunders of the upper deep;
> Far, far beneath in the abysmal sea
> His ancient, dreamless, uninvaded sleep
> The Kraken sleepth . . .

But when the Gulf Stream drives the squid northward into colder waters, the behemoth sea creature rises closer to the surface. Though sightings are rare, there is a documented incident of a giant squid becoming stranded when it swam too close to the Norwegian shore in 1680. Magnus's sea monsters may indeed have been giant squids that surfaced from their deep lairs to wrestle with ships.

The meanders, rings, and eddies of the Gulf Stream represent some of the dynamic forces of nature that not only impact seafarers but also reach the "coastal doorstep" of millions of Americans. Though even landlubbers recognize the ever-changing nature of the ocean, many of them have difficulty in viewing the Gulf Stream from a broader perspective than their human-centered world. Where does the Gulf Stream fit into our often limited view of geography? The folk musician Woody Guthrie in his song *This Land Is Your Land* penned the lyrics "From the redwood forest to the Gulf

Stream waters." Even a couple hundred years before Guthrie, then Secretary of State Thomas Jefferson, in his discussion with European powers on offshore claims, preferred a neutrality zone marked by the Gulf Stream, which he felt was a natural limit. But do the boundaries of this country literally stretch to this current? What ownership, if any, does a nation have over its adjacent waters? What rights of passage do nonhostile nations have on the seaways and currents?

The answers to these questions have a long history, dating back to Roman times, and reflect the reality that much of modern maritime law is based on nationalistic interests. By the middle of the twentieth century, foreign fleets were routinely entering U.S. coastal waters and depleting the fisheries, which had supported generations of cod, haddock, and pollock fishermen. In response to the growing danger of overexploitation of these valuable resources, President Ronald Reagan signed into law on March 10, 1983, the formation of an Exclusive Economic Zone (EEZ), which, in part, excluded all foreign vessels from fishing within a two-hundred-mile zone of the U.S. coast. The law made the Gulf Stream off-limits to all non-U.S. vessels except for navigational purposes. In essence, from an economic, if not geographical, perspective, the Gulf Stream has indeed become the eastern boundary of the United States.

Surface currents and gyres are only one facet of circulation in the oceans. A complete picture of global current flow must include deep ocean circulation. Our view must be three-dimensional and include the coupling between surface and deep flows. While surface currents are wind-driven, deep-water movements are density-driven, depending heavily on changes in water density imposed by air-sea interactions. But why should the surface and deep components of the ocean circulation be coupled? To answer this question, we must accept a very elementary fact of planetary ocean circulation: what goes north must eventually come back south, to maintain equilibrium in the distribution of matter, or mass balance. However, the southward-flowing Canary Current transports only 10 to 15 percent of the water carried northward by the Gulf Stream. What happens to the rest of the water?

In a 1987 article in *Natural History* magazine, Wallace Broecker, a Columbia University geologist, proposed a conceptual model linking the horizontal and vertical components of ocean flow; he dubbed it the "ocean

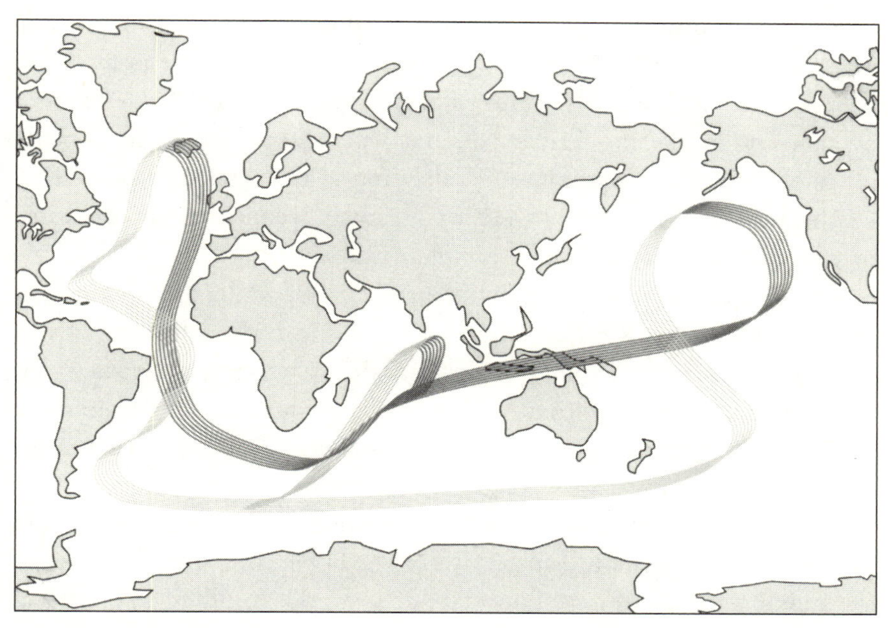

Conveyor-belt circulation, surface component (dark) and deep component (light)

conveyor belt." Though the details of this conveyor belt still need to be worked out, and the original model is somewhat simplified, the essence of its role in water transport can be articulated. Like the conveyor belt in your local grocery store, moving your purchases along to the cashier, there is no true beginning or end to the ocean's conveyor belt. But for the sake of this discussion, the Gulf Stream is as likely a place to begin as any other and represents the upper, or surface, limb of the conveyor. By the time the Gulf Stream reaches the higher latitudes from its journey out of the tropics, its density has increased primarily due to cooling. This dense water sinks into the abyssal depths, reaches a level of equilibrium, and flows southward as the lower limb of the conveyor.

As early as the mid-1950s, oceanographer Henry Stommel (who, as we will see in Chapter 3, would develop a comprehensive picture of gyre dynamics) began thinking seriously about abyssal circulation. One major consequence of his inquiries into the workings of deep circulation was his theoretical prediction of a deep, equatorward current along the North American coast, transporting dense water from the polar to lower latitudes. In 1957, oceanographers Valentine Worthington and John Swallow experimentally verified Stommel's prediction of an undercurrent by tracking metal tubular devices (Swallow floats — named after their developer), which could be made to float at a desired depth. Deploying Swallow floats off to the south of Cape Hatteras, they found the instruments traveled southward at a speed of about a third of a knot — contradicting the commonly held belief at the time of a stagnant abyss.

At Hatteras, Stommel believed the current passed under the Gulf Stream and continued southward into the South Atlantic. Indeed, flowing ever so slowly, this deep-water limb does extend all the way to the southern tip of Africa. Here it joins with other abyssal water and continues its journey eastward into the Indian and Pacific Ocean basins. But in this case, what went down must ultimately come up, and in these basins the deep-water limb eventually surfaces. To complete the conveyor cycle, let's hitch a ride in the Indian Ocean. This surface water, the upper limb, becomes part of the Indian Ocean's western boundary current, the Agulhas, which flows south along Africa's east coast. Rounding the Cape of Good Hope, the upper limb becomes part of an eastern boundary current, the Benguela, which transports water northward through the South Atlantic. There it is picked up by equatorial currents to complete the cycle back to the Gulf Stream — a journey of more than a thousand years. A three-dimensional picture of ocean transport has emerged over time: the brisk flow of water in the surface gyres superimposed upon the slow flow of the abyss.

2 ANATOMY OF THE GULF STREAM

WHILE SOME OUTDOORS enthusiasts have a tendency to overly anthropomorphize every element in nature, it may be unavoidable that one's first impression in encountering the Gulf Stream is how much like a living organism it seems. Similar to the circulatory system of an animal, pumping blood throughout its host, the Gulf Stream unfailingly transports water throughout the western Atlantic. The Gulf Stream waters also have a distinctive coloration and complexion. And the current has a temperature — a relatively high one at that.

Invariably during one of my oceanography lectures I am asked, "Have you *seen* the Gulf Stream?" I answer that I have "seen" it in the best way I know how, leaving the terrestrial world behind and immersing myself in this watery realm.

Your senses soak up the liquid. Shafts of light filter down from above, enveloping you in a sunlit cloak that seems to stretch forever in the transparent water. The supporting water feels warm against your skin; tiny thermals brush up against your body, like bubbles in a hot tub. The stream's saltiness, born of millions of years of continental erosion and volcanic degassing, rests lightly on your tongue. The blood that courses through your body has many of the same salts, primarily sodium chloride, that are dissolved in the stream's flow. You dive deeper; your eyes adjust to the growing dimness. A flash of color passes below. You can barely see the fish; its dark head and upper body, contrasting with its white belly, blend in with the abyss. Countershading affords protection from probing eyes. The water is far from silent, as poets would have you believe. Water transmits sound much more efficiently than air does, and the stream buzzes with all types of racket: from the thunderous explosion of a swift predator breaching the surface in pursuit of fleeing prey, to the subtle clicks, snaps, and crackles of a myriad of tiny organisms. Your body, content to be pulled along by the tug of the current, gently rocks with each wave that passes overhead. The flowing water washes away your anthropocentric past and warms your soul.

Sea surface temperatures in the Gulf Stream are generally above 80°F (27°C), fifteen degrees higher than outside of the current. The Gulf Stream owes its high temperature to the intense sun rays that heat the waters of the tropical Atlantic and Caribbean basins. The Gulf Stream incorporates this heated water via the Antilles, Caribbean, and Loop Currents and transports it northward. The amount transported is not insignificant — on a yearly average, four hundred times more heat than the total consumption of energy in the United States — and plays a critical role in maintaining equilibrium in the earth's climate. Even as the Gulf Stream moves along the southeast coast of the United States, it maintains its thermal characteristics, losing its heat ever so slowly to the overlying atmosphere. The current is a large storage bin of energy. Not only does water cool slowly, it heats slowly as well. A watched pot never boils — or so it is said — reflecting the fact that water requires a relatively large amount of heat to change its temperature even a small amount. Rapid temperature changes over time are not the norm for warm ocean currents like the Gulf Stream. (The reader should be aware that although the words "warm" and "cold" have a strict scientific meaning — referring to heat transfer processes — I will use them in the popular sense.)

As the Gulf Stream approaches the North Carolina coast, the temperature contrast between the coastal Slope Water and the western edge of the stream is more pronounced than along any other portion of its path. This sharp change in temperature over a relatively short distance is a thermal front, analogous to the weather fronts in the atmosphere. From the cold Slope Water (58°F, 14°C) to the warm Gulf Stream, the temperature may increase twenty degrees in only a matter of a few miles. This thermal boundary often marks the confluence of the Gulf Stream and the Labrador Current. While not part of the main Atlantic gyre, the cold Labrador Current flows southward from the Arctic, and a filament of this current contributes to the cold Slope Water. In contrast, the temperature difference between the eastern edge of the Gulf Stream and the Sargasso Sea is not as dramatic as along the current's western edge. The temperature of the Sargasso Sea is a few degrees lower than the Gulf Stream but still considerably higher than the Slope Water. A vessel sailing eastward from Cape Hatteras crosses three distinct thermal regimes: cold Slope Water, a warm Gulf Stream, and a lukewarm Sargasso Sea.

The western edge of the Gulf Stream is where the incessant, flowing, and warm current clashes with the atmosphere to form a meteorological battle field. North Carolina's offshore waters are the breeding ground for a fierce winter storm, called a nor'easter. This is an intense low-pressure area whose center of rotation is just off the coast, and whose leading winds in the left forward quadrant rotate onto land from the northeast. Though nor'easters are extratropical in nature, on very rare occasions — such as the North American blizzard of 2006 and a very severe storm in 1979 — the center of the storm becomes circular, resembling the eye of a hurricane.

Nor'easters owe their formation to the thermal contrast along the Gulf Stream's western edge. When cold, dry Arctic air flows off the continent out over the Gulf Stream, large transfers of heat and water vapor occur from the ocean to the atmosphere. These air-sea exchanges create a very unstable and volatile situation that, when coupled with the energy from the upper-level jet stream, can form a nor'easter. The heat that is released from the Gulf Stream can transform a small storm into a monstrous one, virtually overnight.

On the night of October 31, 1990, an incipient low off North Carolina deepened into a full-blown nor'easter, with winds gusting over ninety miles per hour. This compact powerhouse of a storm ripped a dredge loose from its moorings in Oregon Inlet. Pushed by gale-force winds, the vessel plowed into the Herbert C. Bonner Bridge, causing five spans of the overpass to collapse. Months would pass before North Carolina's Department of Transportation cleared the bridge for the resumption of vehicular traffic. The agency, not to mention the public, was definitely not amused by the pranks of this Halloween storm.

Though this late October storm was localized, many nor'easters travel northward up the East Coast, producing blizzard conditions when they begin to tap the Atlantic's moisture supply. Wind-driven waves, generated by storms with gale-force winds, then batter the coast from Virginia to Maine, resulting in flooding and severe beach erosion.

The "Blizzard of 1996" dumped record amounts of snowfall from the Carolinas through New Jersey. The Appalachians were particularly hard hit, with snowfall totals of more than three feet. The storm formed in the Gulf of Mexico on the morning of January 6, 1996, moved south of Alabama, and reached Savannah, Georgia, the morning of the seventh. Now over the Gulf Stream, the storm tracked almost due north, reaching Cape Hatteras that evening.

A classic setup for the explosive development of a nor'easter was in place. A cold dome of high pressure was over southern Canada and New England. The importance of the high is twofold. First, it serves to impede the northward progression of the deepening storm, resulting in a long-duration precipitation event. Second, the high funnels cold air from New England toward the Appalachians. Since the air is too dense to flow over the high terrain, it dams up against the mountains, and a large pool of frigid air settles over the mid-Atlantic region. Temperatures on January 7 in the affected area hovered in the twenties.

As the jet stream sped across the Hatteras low, the pressure began to fall, surface air was swept aloft, and the low intensified. A meteorological chain reaction was now beginning. The low began to pull in the cold air from the coast to replace the air forced aloft. As this cold air mass glided over the 65° waters of the Gulf Stream, massive amounts of sensible and latent heat, released from the condensation of water vapor, poured into the atmosphere. This coastal storm, fueled by very warm, vapor-laden air over the Gulf Stream and strengthened by a powerful jet stream, metamorphosed into a monster storm by nightfall of January 7. Overnight, gale-force winds buffeted the coast, fifteen- to twenty-foot waves ate away at the beaches, and snow blanketed the eastern seaboard. By January 8, the storm had departed the mid-Atlantic region, leaving in its wake millions of residents with the backbreaking task of digging out from mountains of snow.

Meteorologists refer to these very intense nor'easters as "weather bombs," a term stemming from World War II. The analogy is twofold: their explosive nature and their sudden appearance. Over the years, the term evolved into a description of an extratropical cyclone with a specific definable and measurable characteristic: the core barometric pressure drops a minimum of twenty-four millibars (seven-tenths of an inch of mercury) in twenty-four hours. (Some of these systems have experienced a decrease of pressure of more than sixty millibars.)

These storms can generate hurricane-force winds of more than seventy-four miles per hour. A "bomb," churning over the warm waters of the Gulf Stream, can easily form steep, sharp-crested waves fifty feet high — a major threat to ships at sea. Even with the technology available today, prediction of these rapidly deepening "bombs" often proves elusive; a fact dramatized in the book *The Perfect Storm*. What chance did the unsuspecting slave traders, pirates, pilots, merchants, and colonists of the sixteenth and seventeenth centuries have of surviving the Gulf Stream's reputation as the "weather maker" of the western Atlantic?

Timely predictions of nor'easters depend, in part, upon the location of the Gulf Stream's thermal fronts. But they are not fixed in space and time. Pronounced meanders may significantly shift the location of these fronts. As a meander forms to the north of the prevailing path of the Gulf Stream, warm Sargasso Sea water intrudes into the Slope Water. If the meander is pinched off from the main flow, then as we have seen, a ring will form to the north of the Gulf Stream. But since such a ring will have incorporated the thermal characteristics of the Sargasso Sea, it will be a warm-core ring, surrounded by the colder Slope Water. Conversely, a southerly ring will have a cold core that is embedded in the Sargasso Sea. Decaying slowly, these rings gradually lose their distinct thermal characteristics and ultimately become indistinguishable from their environment.

Though the Gulf Stream is reluctant to give up its heat on its northward journey, heat is ultimately lost to the atmosphere at the higher latitudes. Similar to the way blowing on a spoon of hot soup speeds up the cooling process, the strong westerly winds, blowing off the cool Canadian landmass, accelerate the transfer of heat between the sea surface and the atmosphere. By the time the Gulf Stream reaches the middle of the North Atlantic, it has cooled significantly from massive amounts of heat pouring into the overlying atmosphere. With a temperature drop of more than thirty degrees, the Gulf Stream loses one of its most defining and distinguishable characteristics as it continues its eastward journey across the Atlantic.

As early as 1606, the French historian Marc Lescarbot recognized that the key element in assessing the position of the Gulf Stream was the warmth of its water. A lawyer by training, but an explorer at heart, Lescarbot recorded the following observation while aboard the ship *Jonas*: "Six times twenty leagues to the eastward of the Banks of Newfoundland, we found for the space of three days the water very warm, whilst the air was cold as before, but on June 21 quite suddenly we were surrounded by fogs and cold that we thought to be in the month of January, and sea was extremely cold." While Lescarbot erroneously attributed the cold sea temperatures to the presence of ice that came floating "down from the coast and sea adjoining to Newfoundland and Labrador, which is brought thither by the sea in her natural motion," he was observing, unknowingly, the clash of opposing ocean currents. At these latitudes, the Gulf Stream again meets the Labrador Current.

The use of temperature measurements as a diagnostic tool in assessing the nature of ocean currents continued on into the American Revolutionary period. In addition to being a statesman and diplomat, Benjamin Franklin was also a scientist. While probably best known for his experiments with lightning, Franklin became intrigued by the idea of a "stream" flowing through a large body of water such as the Atlantic Ocean. In part to satisfy his curiosity and driven by the economic necessity to improve mail routes to Europe, Franklin took water temperature measurements on three North Atlantic crossings and meticulously recorded them. From these readings, he was able to determine the position of a vessel in relation to the Gulf Stream and even how far the vessel was from shore. Franklin clearly expresses his findings about the thermal characteristics of currents: "I find that it [the Gulf Stream] is always warmer than the sea on each side of it. I annex hereto the observations made in two voyages and may possibly add a third. It will appear from them that the thermometer may be a useful instrument to the navigator, since currents coming from the northern into the southern seas, will probably be found colder than the water of those seas as the currents from southern seas into northern are apt to be warmer." In typical Franklin fashion, his observations were quite extensive — even including on his last voyage in 1785 the first attempt to measure subsurface temperatures — and generated considerable interest within the scientific community.

One of the most prominent disciples of accurate temperature measurements was Dr. Charles Blagden, who while with the British fleet sailing American waters in 1776–77, observed the temperature in the Gulf Stream off Cape Fear and also off Chesapeake Bay. In a letter to the Royal Society in 1781 communicating his findings, he strongly expressed his opinions regarding the nautical advantage gained by the use of the thermometer. Though Franklin and Blagden were the first to demonstrate the usefulness of this instrument in marine navigation, their work was just the beginning of thermometry in the systematic study of ocean currents.

On Franklin's last voyage he was accompanied by his nephew, Jonathan Williams, who proved to be an able assistant to Franklin in his measurements and recordings. Franklin awakened such an interest in his young protégé that Williams enthusiastically continued the experiments begun by his uncle. His research culminated in the 1799 publication of *Thermometrical Navigation*, which included a map of the Gulf Stream based upon his thermal measurements. In its time, this manuscript was the navigational standard-bearer for mariners.

In order to test Williams's results, a Captain William Strickland undertook numerous voyages across the Atlantic during which he performed daily and often hourly observations of sea surface temperature. From this extensive set of measurements, he was able to delineate the relatively warm northeast branch of the Gulf Stream as it flows away from the North American mainland. Though others before him had noticed signs (floating weeds and debris) of a current flow at these high latitudes, and even the fifteenth-century explorer John Cabot remarked that the beer in the hold of his vessel was getting warm, Strickland was the first to confirm the existence of this current.

One of the more interesting findings from these early thermometric measurements would come from the cruise of the vessel *Eliza* while en route from Halifax to England in April 1810. Though positioned near the warm water of the Gulf Stream, the crew of the ship discovered a large mass of cold water, two hundred miles in diameter, estimated to be ten to fifteen degrees colder than the surrounding water. The scientific personnel aboard the vessel proposed this pool of cold water was the result of icebergs that became entrapped and melted in the Gulf Stream. With the advantage of our present-day understanding of the current, we can surmise that this was the first direct observation of a cold-water ring.

While the Age of Discovery (mid- to late fifteenth century) marked a period of "science for voyaging" — a growing understanding of the ocean itself for the sake of exploration — the eighteenth century ushered in the era of "voyaging for science." By the beginning of the nineteenth century, this movement was in full swing, and the study of ocean currents was the focus of many investigators. With the simple thermometer still the instrument of choice, scientists and mariners intensively probed and examined the Gulf Stream. In particular, the published works of Alexander von Humboldt in 1814 and Captain John Hamilton in 1825 showed that the location of the Gulf Stream exhibited seasonal variability as a result of changes in atmospheric flow. These findings were hard-won; Humboldt and Hamilton crossed the Gulf Stream and the North Atlantic no less than sixteen and twenty-six times, respectively.

In the years to follow, temperature measurements would continue to add to the growing body of knowledge about the Gulf Stream and all of its nuances. But the process of taking the stream's temperature would become more sophisticated with advances in instrumentation. While the first thermal measurements simply involved sticking a thermometer in a bucket of seawater, the analytical techniques of today allow for the data to be transmitted electronically to the ship, where the onboard scientists continuously monitor spatial and temporal temperature changes.

The greatest advancement to our understanding of ocean currents, including their thermal characteristics, has come from remote sensing platforms; in particular, satellites. Instruments aboard satellites orbiting hundreds to thousands of miles above the earth's surface produce images of wide swaths of the ocean surface — essentially providing the viewer a complete portrait of the phenomenon in question. While visible images, similar to the pictures a photographer might record, yield some information about the nature of an ocean current, their value is intrinsically limited. In contrast, infrared images display thermal gradients, or differences in surface temperature over a distance, by coding the amount of heat emitted from the surface into definable temperatures.

While even the lay community, with a little training, can use infrared images in their unadulterated state to determine temperature, the fishing industry routinely uses them to locate fish. Satellite images are broadcast directly to fishing vessels to help guide them to the most favorable areas,

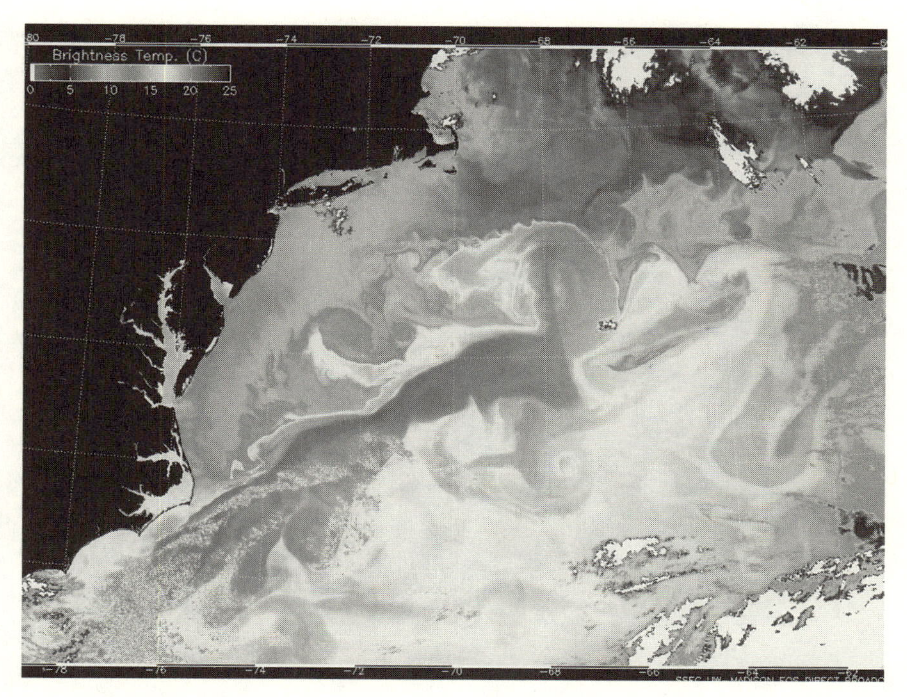

Infrared image of the Gulf Stream departing the North Carolina coast (courtesy of NASA)

thermal fronts and rings, holding fish. Though Benjamin Franklin would be in awe of the sophisticated technology available today to gather data, the underlying principle remains unchanged since his time: thermal differences delineate the Gulf Stream's rings, loops, and splits.

Some within the general populace have long held the belief that the Gulf Stream warms Europe and accounts for its relatively mild climate throughout the year. Acceptance of this role of the Gulf Stream is so widespread that it has become folklore in some academic circles. This idea may date back to Matthew Maury, who waxed eloquent about how the Gulf Stream "makes Erin the 'Emerald Isle of the Sea,' and clothes the shores of Albion, in evergreen robes, while in the same latitude, on this side, the coasts of Labrador are fast bound in fetters of ice."

Not true, argues a group of climate scientists headed by Richard Seager of Columbia University's Lamont-Doherty Earth Observatory, whose research has shown that the moderating influence of this current is over-

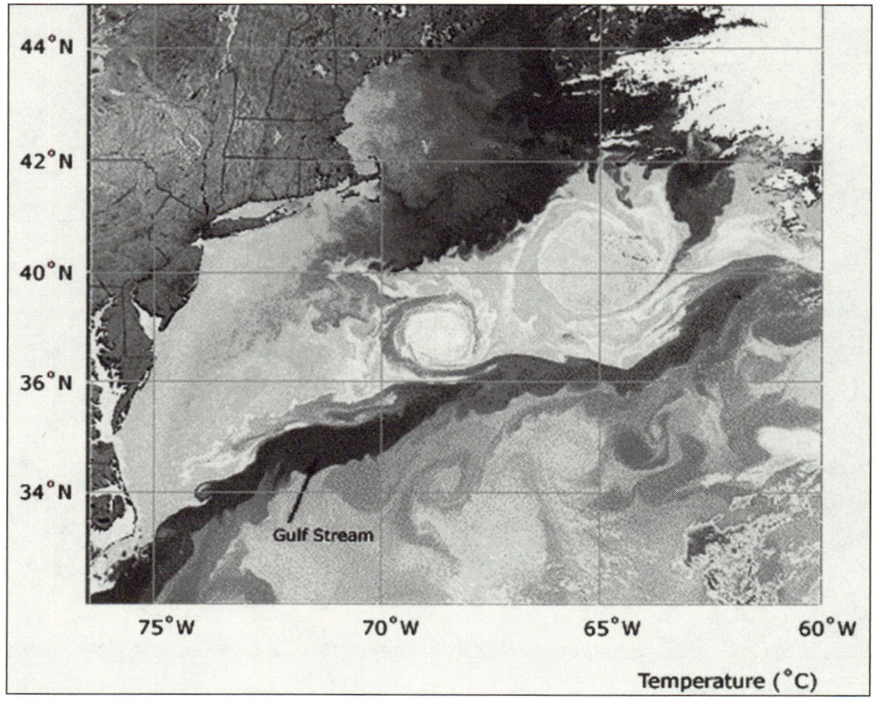

Warm-core rings, north of the Gulf Stream (courtesy of NASA)

stated. The Gulf Stream does indeed transport heat from the tropics to the shores of England, but only about 10 percent of England's winter warming comes *directly* from the Gulf Stream. Westerly winds, blowing across the long fetch of the Atlantic Ocean, account for the remainder of the warming by picking up heat from the relatively warm sea surface. The vast expanse of the North Atlantic is like a giant oceanic vault, retaining heat until the overlying winds make a withdrawal. The amount transferred can be large. Marine observations fed into a computer showed that 80 percent of the heat that warms England is from this source. These same westerly winds, blowing over the Gulf Stream, also extract some of its heat — a contribution of the final 10 percent to the warming of England and beyond.

Though the Gulf Stream derives much of its renown from being the main artery in the North Atlantic carrying warm water throughout the hemisphere, our picture of it would be incomplete if we viewed it solely from the narrow perspective of temperature. For probably nothing catches

Weakening of the Gulf Stream at higher latitudes (courtesy of NASA)

our attention more than a vivid landscape or an ocean view. We photograph, paint, and even record these vistas in our patriotic songs: "For amber waves of grain / For purple mountain majesties." We are enamored with the colors of nature, be it a blood-red sunset or the dark green canopy of a Maine forest.

While our perceptions of the environment today are perhaps dulled by our dependence on instrumentation, the early seafarers were keen observers of their environment since their lives often depended on the successful reading and interpretation of their surroundings. And the Polynesians were probably at the pinnacle of all navigators in developing acute powers of observation. Their navigation feats date back thousands of years, and the color of water became one of their key navigation tools. For instance, they could interpret the green of sunlit lagoons reflected off the bottom of clouds as an indication of faraway coral atolls.

The earliest known recorded account of the visual attributes of the Gulf

Winslow Homer's *The Gulf Stream* (The Metropolitan Museum of Art, Catharine Lorillard Wolfe Collection, Wolfe Fund, 1906 [06.1234]. Image © The Metropolitan Museum of Art.)

Stream probably dates back again to Maury, who observes: "Its waters, as far out from the Gulf as the Carolina coasts, are of an indigo blue. They are so distinctly marked that their line of junction with the common sea-water may be traced by the eye."

Winslow Homer, one of America's premier painters of seascapes, was intimately familiar with Maury's discourse on the Gulf Stream, and he claimed to have personally crossed the Gulf Stream no less than ten times. These collective experiences led to one of his best-known works, *The Gulf Stream* (1899) — a harrowing scene of a seemingly dazed black man perilously adrift in a derelict boat surrounded by sharks. While the interpretation of this work has generated considerable debate among art historians, most acknowledge that it is a realistic portrayal of this blue river and recognize him for his keen and straightforward observations of nature.

While satellite technology has opened new opportunities to study the Gulf Stream, the images, at least to me, are sterile, lacking any personal connection to the current. I had often visualized how it might look from the perspective of a sea bird, with its outstretched wings, gliding languidly over the current. The closest I would come to this type of flight would be in a small, single-engine plane. When I explained my desire to fly over the Gulf Stream to Dwayne, proprietor of an air tour service on the Outer Banks and a longtime Banker, he incredulously replied, "Why do you want to go there? It's only water. Don't you want to see the islands, villages, and lighthouses?" After further discussion, which still did not completely convince Dwayne, we agreed upon a charter for the next day.

After clearing the airport runway, we banked east, heading to the Gulf Stream. As the plane passed over Cape Point, the easternmost tip of Hatteras Island, the cocoa-colored waters along the coast came markedly into view — definitely not a picture postcard setting. The Bermuda high, which is anchored off the North Carolina coast during the summer, is responsible, in part, for this drab coloration. On the western side of this huge pressure center, persistent southwesterly winds stir up shallow water sediments, discoloring the water. The sea surface was so roiled up that I doubt the bathers below, standing waist deep in the water, could see their toes.

As the plane's small engine whined on like a lawn mower, a subtle transformation in the water's color became apparent, the brown giving way to more shades of green and blue. The blue river was drawing nearer. Then the line appeared; the same color line Maury eloquently wrote about. It was so distinct that I could almost envision Neptune ascending from his watery kingdom, marking the boundary of his domain. The blue of the sea, which stretched as far as the eye could see, meshed with the blue of the sky on the horizon. I absorbed it all.

But I soon found myself in a precarious position, or at least I felt that way, when Dwayne pushed the plane's stick forward to, as he explained, "Get a better look at the current." The nose dropped, and my stomach lurched up. We skimmed over the white caps, which had popped up on the blue veneer, and followed the color line for miles, observing that the line is not laser straight but, like a flag flapping in the breeze, undulates under the push and pull of wind and sea. With the tour winding down, I felt quite fortunate to have seen the Gulf Stream from a perspective that Benjamin Franklin and Winslow Homer could only have imagined. Even a converted Dwayne couldn't mask his enthusiasm, blurting out, "Why is the Gulf Stream blue?"

The Gulf Stream owes its coloration to its complexion — markedly clear, devoid of any significant amounts of sediment or plant material. During our flight, the strong summer sun had afforded Dwayne and me the opportunity to peer into the depths of the stream. The water was so clear and transparent that both of us had the strange sensation of peeping into a hidden, private place.

Most tropical waters are nutrient-poor and cannot support much plant life. Nutrients like fertilizers are essential elements needed by plants to encourage tissue growth. The major nutrients, nitrates and phosphates, are released into the marine environment through microbial processes often occurring on or within the seafloor. And here is the conundrum: While marine plants need nutrients, they also require the sunlight found at shallow depths for photosynthesis. Thus, the plants are physically separated from the abundant cache of nutrients located in the deep ocean realm. So the dominoes begin to fall: no nutrients, no plants, crystalline blue water. But what controls this pronounced separation of plants and nutrients throughout their watery realm?

The answer to this question can be found by considering a fundamental feature of all the world's oceans: deep water is perpetually cold. While this fact is certainly not news to most people in the twenty-first century, it was a major revelation to eighteenth-century Europeans. In 1751, Captain Henry Ellis, piloting an English slave ship in the subtropical Atlantic, made a significant discovery. Reverend Stephen Hales had supplied to Ellis a unique instrument, consisting of a system of valves, which could sample water from various depths when lowered by means of a rope, and the temperature of these samples could be determined from a built-in thermometer, when the instrument was retrieved and brought aboard. To his surprise, the inquisitive captain discovered that the deep water was quite chilly. In this region, the warm surface waters are mixed vertically by winds and waves to form a relatively shallow (three-hundred-foot deep) layer of uniformly warm water. Below this isothermal layer is the thermocline, which is a zone of rapidly decreasing temperature with depth. Extending down to a depth of approximately three thousand feet and bottoming out with a chilly temperature of about 40°F (4°C), the thermocline is a transition layer separating the warm surface layer from the cold deep water. Captain Ellis was the first to recognize this thermal transition, which he reported in a letter to a Reverend Hayes: "The cold increases regularly, in proportion to the depths, till it descended to 3900 feet: from whence the mercury in the thermometer came up to 53 degrees; and tho' I afterwards sunk it to the depth of 5346 feet, that is a mile and 66 feet, it came no lower." In most cases, these strata do not mix readily since warm water is less dense than cold water. Separated by their respective densities, the nutrient-enriched deep water is essentially isolated from the surface.

Since these layers of the ocean are not prone to overturning, the Gulf Stream conveys this nutrient-poor water hundreds of miles from its source. Though the analogy of the Gulf Stream as a desert with regard to the dearth of plant matter is not farfetched for the most part, there are oases of life in the current. These fertile spots occur when nutrients from the bottom are made available to surface-dwelling plants. For example, as the Gulf Stream encounters the aforementioned, relatively shallow Charleston Bump, deeper waters of the Gulf Stream upwell and infuse the sunlit layers with nutrients.

The frontal boundary separating the Gulf Stream from the Slope Water also marks a transition in nutrient concentration. The Slope Water exhibits marked fertility because of the nutrient-enriched water from the north. As the meanders of the Gulf Stream begin to entrap this cold, nutrient-rich water, cold-water rings complete the process and laterally transport these nutrients to the barren Sargasso Sea.

The great depth of the Sargasso Sea, more than fifteen thousand feet in places, is responsible for the lack of nutrients and thus microscopic plant matter within its surface waters. Bottom nutrients are literally entombed in this deep watery grave, and the surface waters are bare. The warm-core rings, which originate in the Sargasso Sea and eventually end up in the Slope Water, have entrapped this nutrient-poor surface water. Whether rings are warm or cold, nutrient-rich or nutrient-poor, they are the Gulf Stream's only mechanism for the exchange of water across its boundaries.

If not for rings, the Gulf Stream would be a watery barrier, permanently isolating Slope Water from the vast Sargasso Sea.

If we now view the Gulf Stream as the main artery of the Atlantic transporting water, nutrients, and heat in a timeless process, then we must look for its heart. What pumps this powerful flow of water? Where is the source of its power?

3 FLOWING DOWN THE HILL

THE HISTORY OF OCEAN CIRCULATION

BY THE NINETEENTH CENTURY, detailed charts depicted the main patterns of ocean surface circulation with a relatively high degree of accuracy. Though these charts proved to be invaluable to the maritime community, what was missing was a scientifically grounded explanation of this unique circular flow. To many of the learned, the complex of Atlantic currents remained one of the most intriguing and challenging natural systems of their time. What caused the gyres? What was the relationship between air and water currents? Ideas and speculations flowed from many segments of society, including scientists, philosophers, and the clergy. But a lack of understanding of the earth's geography, human tendencies toward hyperbole, and the influence of religious dogma led to many implausible theories being put forward that ultimately had to be discarded. It would not be until the next century, with more than a few wrong turns along the way, that science would arrive at a unified theory of gyre dynamics, including how the Gulf Stream fits into this picture.

The Greek poet Homer (ca. 850 B.C.), famous for the poetic epics the *Iliad* and the *Odyssey*, envisaged the Earth as a giant disk with the ocean flowing like a river around it. This idea of a moving stream of ocean is repeatedly taken up in the *Odyssey*; for example, after a day's voyage, Odysseus and his men were said to have "attained Earth's verge and its girdling river of Ocean." But what was Homer's basis for his view of an ocean-river? Evidence suggests that this picture of the Earth may have even predated Homer, and that by his time, the Greeks and other Mediterranean inhabitants simply accepted the ocean-river as fact. No plausible explanation was forthcoming from Homer or from those to follow as to why the ocean should flow like a river. But through conjecture and legend, it appears, the Greeks were willing to accept the view of the known land of their time being surrounded by an ocean. Having long been seafarers and obviously aware of coastal currents in the Mediterranean, the Greeks may have developed something like the following argument: where there is a sea there are currents, and if the land is bordered on all sides by an ocean, then a riverlike flow would be found girdling the disk. This belief would continue into the Greek Classical Age. And although some refinements were made to it, like the idea of Pythagoras (ca. 550 B.C.) that the Earth was a sphere instead of a disk, geographers and historians of that period were generally in agreement that an ocean-river circled the earth. Herodotus in 450 B.C. compiled a map showing the Mediterranean Sea surrounded by the landmasses then known as Europa, Asia, and Libya. Large tracts of ocean, which the Greeks believed extended to the very edges of the world, surrounded this large landmass. Though many learned men of this period avoided addressing the question of the ocean-river phenomenon, lacking the hard evidence to provide any answers, philosophers dove head first into the murky waters of this inscrutable mystery. They were more than willing to provide a host of possible explanations, including many that would not stand up to scrutiny.

The philosopher Socrates, who lived from about 470 to 399 B.C., was the first to propose, incorrectly, that the ocean currents were the result of deep subterranean channels in the Earth transporting huge volumes of seawater. He proclaimed that within the Earth there were many deep cavities connected by great flowing streams, the greatest of which is Oceanus. In Greek mythology, Oceanus was the unending stream of water encircling the Earth, a Titan whom the Greeks viewed as the origin of all things on Earth. Homer in the *Odyssey*, for instance, cautions that this great river must be crossed to reach the Underworld. As the god of the sea, Oceanus was both respected and feared by the ancient Greeks. Who would be foolhardy enough to challenge this deity?

Aristotle (384–322 B.C.) would later go on to reject many of the ideas about the oceans emanating from the Greek Classical Age, including the Earth-encircling river of ocean. But Aristotle discussed the concept of motion at great length, and he was ultimately to come to the conclusion that all movement is derived from a divine force, the *primum mobile* (prime mover), forever propelling the heavens toward the west. While he offered little direct explanation regarding the motion of the sea, his treatise on motion would heavily influence the way humankind perceived ocean circulation long after his time. Centuries would pass before any significant new theories about ocean circulation were proposed.

Following the Dark Ages (476 to 1100), a period in Europe characterized by many historians as one of intellectual stagnation and widespread ignorance, early Renaissance scholars continued to believe that oceans should move from east to west, as Aristotle's idea suggested. They simply accepted as fact, without question, the *primum mobile*, even though the perplexing issue of where the water went once it reached the other side of the Earth remained unresolved. If the water neither piles up in the west nor empties in the east, what becomes of it?

One of the first attempts to resolve this question was a discourse by Peter Martyr in 1515. Though an advocate of the general principle of westward motion, he proposed that water flows in the form of a current that ultimately circulates around the planet. His idea of a current and ocean circulation, while accepted today, was very insightful for the sixteenth century and would not be grasped by many others in the following 150 years. But for his hypothesis to be true there had to be an outlet for the current after it reached the western side of the Atlantic. In his writings, Martyr attempted to address this sticking point. He discussed the commonly held opinion of there being an outlet from the Atlantic to the Pacific through the Americas. But he categorically rejected this idea because voyages starting with Columbus and continuing into the sixteenth century had produced no evidence for such an egress. (Martyr must have been intimately familiar with the exploits of the early Spanish explorers, since as the Spanish royal chronicler, he wrote some of the first historical accounts of the voyages of Columbus, Ojeda, and Pinzón.) Instead, Martyr was ultimately to come to the belief, which was firmly grounded in Aristotelian philosophy, of a large passage through the Canadian north. Though his conviction would ultimately prove to be erroneous, it would, in the future, bolster arguments by the British in their search for a Northwest Passage to the riches of the Orient. Unfortunately for the British, their intense desire to find a path to the East that was outside the monopolistic realms of the Spanish and the Portuguese would color many of their forays into the theories of ocean circulation.

Voyages by the Portuguese and Spanish fleets during the mid- to late fifteenth century brought unprecedented opportunities to make observations of many of the Earth's most important ocean currents. In particular, the men who sailed for these nations would become intimately familiar with the ocean's equatorial and boundary currents and subsequently utilize these currents to expand their nation's economic ventures. They discovered that some currents, like the North Equatorial, flowed to the west in the direction dictated by the *primum mobile*—an idea widely championed by theologians and philosophers. But others, like the North Atlantic Current, flowed against the grain of religious and Aristotelian expectations of westward movement. With the rapid accumulation of knowledge regarding ocean circulation during the fifteenth and sixteenth centuries, it became increasingly difficult for scientists to reconcile this new information with accepted doctrine. By the year 1519, the Spanish were routinely sailing along with currents to America and then northward along the coast. But some theological circles steadfastly refused to recognize the returning, eastward flow of the northern limb of the North Atlantic gyre. The conviction of only westward motion, according to the *primum mobile*, would not allow for such an admission.

Once seafaring nations accepted that a tropical passage connecting the Atlantic and Pacific Oceans did not exist, they began to modify the concept that only westward motion was possible, but the *primum mobile* would still remain as a central tenet of oceanic circulation. The discovery of westward-flowing equatorial currents in the Pacific and Indian Oceans would reinforce the arguments for this ancient concept.

The British navigator Sir Humphrey Gilbert (1537–83) would champion Martyr's belief that water would have to flow northward along the American coast before flowing again westward through a Canadian passage. Gilbert was so convinced of the existence of a Northwest Passage that he argued vehemently for its existence based upon a continual flow along the coasts of North and South America. While implicitly recognizing the Gulf Stream as the conduit transporting water northward, he doubted the existence of its eastward extension, still clinging to the archaic idea of a general westward motion. But Gilbert's conviction in what he believed to be the truth about ocean circulation may have precipitated the first attempt to describe the nature of the Gulf Stream.

André Thevet (1502–90), a cosmographer to King Henry III of France, added a new wrinkle to the westward-flow concept, arguing that in lower latitudes this flow strikes the land where it reaches its highest level, before turning north as a strong downward-moving current, analogous to water flowing down a hill. (As we will see, this hill analogy would again be used in the twentieth century, but in a different format.) Probably not totally comfortable with his initial explanation, Thevet added that fresh water discharged from rivers into the Gulf of Mexico could also contribute to the surge in the Gulf Stream flow as it travels past Florida on its northward journey. Thevet's suggestion of the importance of river input to ocean circulation is not supported by modern deep-sea soundings, which show an ocean so deep that it seems unlikely whether even the mightiest rivers could contribute more than a small amount of water to it.

The pervasiveness of the belief in a general westward flow was so woven into the fabric of late sixteenth-century societies that believers routinely dismissed as heresy even reliable observations of counterflows. Though the Age of Discovery had brought new information about the nature of winds and currents, the theories based on observational evidence that were needed to explain these phenomena had not changed significantly since Aristotle. In comparison to the blossoming of scientific understanding in astronomy, anatomy, and botany during the sixteenth century, progress in the atmospheric and oceanic sciences was simply not as rapid. Cavalier attitudes toward the role of observational evidence in understanding and explaining fluid flow would soon start to fade, albeit slowly, as the Renaissance period came into full bloom.

In the mid-seventeenth century, Bernhard Varen (some references have latinized the name to Varenius) summarized the current state of knowledge regarding the oceans and their currents in the widely referenced book *Geographia Generalis*. Published at the time of his death, it was the first comprehensive, objective compilation of knowledge about the Earth's geography and would distance itself from the *primum mobile* view of ocean circulation. In it Varen gave a wide-ranging discourse on the nature of winds and currents and classified ocean currents as either steady or periodic. He placed the oceans' gyres in the category of "perpetual special" motions and described the Gulf Stream as a gigantic stream, beginning at the eastern cape of Brazil and flowing northward to its terminus in Florida. Varen was probably the first to show a relationship between the winds and these steady currents, but those who came after him would have to develop the details of this connection. Undoubtedly, Varen's major contribution to the advancement of ocean science was to force philosophers to reconcile their theories with the growing knowledge of the sea.

Human understanding of oceanic circulation progressed rapidly during the seventeenth century. Isaac Vos (or Vossius) published in 1663 *A Treatise concerning the Motion of the Sea and Winds*, which became the seminal work on the subject during this period. In the book, the Dutch librarian Vos presented the first "big picture" of ocean circulation, explicitly describing the nature of the oceanic gyres. His model begins with westward flow in the tropics, which, he argues, if not for the obstruction provided by land, would flow continuously around the globe (*primum mobile*?). Like Varen, Vos assumed that all other currents resulted from a combination of tropical flow and landmasses, which comprised a closed circulation within the ocean basins. His description of the North Atlantic gyre is uncanny in its accuracy, considering it was written more than three hundred years ago: "With the general equatorial current, the waters run toward Brazil, along Guyana, and enter the Gulf of Mexico. From there, turning obliquely, they pass rapidly through the Straits of Bahama. On the one side they bathe the coasts of Florida and Virginia and the entire shore of North America, and on the other side they run directly east until they reach the opposite shores of Europe and Africa; from thence they run again south and join the first movement to the west, perpetually turning in this manner circuitously."

Vos's detailed description of closed oceanic gyres did not receive widespread acceptance among scholars, particularly those with a strong religious orientation. Their views can be traced to the sentiments that were to come out of the Counter-Reformation, a period of Catholic revival during the sixteenth and seventeenth centuries. Again, religious dogma and fervor held sway over scientific reasoning. However, science writers who embraced mainstream theology found it increasingly difficult to resolve some of their beliefs in light of the observational evidence.

In 1665, Athanasius Kircher, a Jesuit priest, produced the first published chart of global ocean circulation. Unfortunately, Kircher's belief in Aristotle's concept of the *primum mobile* severely limited his interpretation, and his map shows extensive westward flow nearly filling the basins of the Pacific and Indian Oceans. Even more problematic are his detailed depictions of entrances to a vast system of subterranean channels that would facilitate the general westward motion. However, falling back on his scientific instincts, Kircher does a better job in accurately depicting the circulation in the intensively studied Atlantic Ocean. In particular, his chart was the first showing the location of the Gulf Stream. Since he was a conservative cleric and scholar, his map ultimately reflects his difficulty in resolving and maintaining his belief in classical Greek ideas in light of expanding knowledge about the oceans and a move toward empiricism.

Another map, published by Eberhard Happelius in 1685, would again show the Gulf Stream in its proper geographical framework, but this map and Kircher's were published more for scientists than for the practical benefit of mariners and traders. Since the public sector did not openly embrace these maps, many maritime historians view them as nothing more than a footnote in the chronicles of ocean circulation.

By the late 1600s, only the Dutch, who by this time had become a leading mercantile nation, were publishing sailing directions for their trading fleets. But, as with the charts of Kircher and Happelius, the at-large maritime community found little value in these maps; they were localized in scope and contained no information on sea-lanes in the open ocean. It would be left to William Dampier to address this issue. As an English explorer, naturalist, and even buccaneer, Dampier traveled extensively throughout the world's oceans and made meticulous observations of winds and currents. From his accounts of these journeys and adventures, he published in 1699 his *Discourse of the Trade Winds, Breezes, Storms, Seasons of the Year, Tides, and Currents of the Torrid Zone throughout the World*, which would establish his reputation as a hydrographer beyond challenge. He differentiated between the various types of winds, such as seasonal monsoons and the steady trade winds and provided detailed maps of them in his discourse. He established that the region of equatorial currents coincided with the trade winds, and he went further than Varen in being the first to deduce correctly from his observations that these winds were the driving force behind the equatorial currents. For the first time in history, Dampier had set forth a realistic description of wind-current coupling to the scientific community. They would finally have to stand up and take notice.

But what could historians conclude about humanity's understanding of the intricacies of wind-current interaction at the end of the seventeenth century? How much had humankind progressed from the theories of Socrates and Aristotle? The marine historian Harold Burstyn makes a persuasive case that, in spite of the growth in maritime commerce, the increase in accuracy of ocean measurements, and the falling out of favor of long-held concepts, wind and ocean current theories still lacked the degree of sophistication that was needed to explain the connection between these fluids. Even though Isaac Newton had developed the science of moving bodies, and this information was codified in his *Principia* published in 1686, the application of Newtonian physics to the motion of the atmosphere and the ocean, particularly the Gulf Stream, would have to wait until the twentieth century.

In the two centuries following the initial Spanish encounter with the Gulf Stream off the Florida coast, European nations expended very little effort studying the Gulf Stream in detail. This might seem surprising in light of the long-established trading and slave routes between Europe and North America, a history that will be taken up later in this book. It is possible that some countries undertook systematic studies of, for example, current flow and structure, but most likely such information was shrouded in secrecy to maintain an economic advantage. A slaver of this period who was in possession of a detailed map of the Gulf Stream depicting specific navigational directions and sea routes would have had a distinct advantage over his competitors because he could shorten his sailing time. One can only imagine the consequences he would face from his employer if such a map fell into the hands of his enemy, or worse, pirates. A ship's captain would have guarded these maps like gold, and when in peril, he would have scuttled them — to be lost forever.

Though the Gulf Stream gained little public notoriety from world traders, even less was known, or at least publicized, about the eastward extension of the Gulf Stream, which merges into the North Atlantic Current. Could it be that remnants of Aristotle's concept of westward flow throughout the hemisphere still seeped into eighteenth-century thinking and prevented the widespread dissemination of information on any counterflow?

The New England whaling community stood out for its knowledge of the eastward extension of the Gulf Stream. In their ever-expanding search for better whaling grounds, the whalers soon discovered the propensity of whales to congregate near the edges of the Gulf Stream. American whalers, by necessity, developed intimate knowledge of this current's flow, variability, and structure, which ultimately led to a fortunate twist in the fate of the American colonies. Their knowledge would help to shape the legacy of one of America's most illustrious statesmen, Benjamin Franklin.

In 1768, English authorities queried Franklin, then the deputy postmaster general for the American colonies, as to why it took two weeks longer for the mail ships to sail between England and the colonies than the merchant ships. Perplexed, Franklin consulted his cousin, Timothy Folger, a whaling ship captain from Nantucket. Folger informed Franklin that the mail ships were unknowingly sailing against the current. In contrast, the captains of the merchant ships knew to stay out of the current on their voyage to the colonies. Folger also related that the whalers would often cross the Gulf Stream to meet the struggling mail ships and attempt, usually unsuccessfully, to convince them to move out of the current. Armed with this information, Franklin had a chart of the Gulf Stream printed in London in 1769. It included detailed instructions by Folger on how to avoid the current when sailing from Europe to North America. In addition, the chart, considering the time of its construction, was an uncanny depiction of the nature of the Gulf Stream. The average path, boundaries, width, and speed of this current, as we know it today, are accurately represented on the Franklin-Folger chart. But for reasons not totally clear, possibly because of growing revolutionary tensions between England and her American colonies, the chart did not meet with wide acceptance from English shipmasters and went rapidly into obscurity. It was not until 1786, with the third printing, that the chart stirred considerable interest among hydrographers and even the general public, propelling the name "Gulf Stream" into common use. Once politicians recognized this mighty river played a significant role in navigation, all government vessels were required to routinely take observations when traveling along its path.

While the forerunner Gulf Stream maps by Kircher and Happelius have faded into relative obscurity, the Franklin-Folger chart succeeded, in part, because of its relevance to the fledgling American colonies. In today's world of almost instant communication, the idea of a letter taking two weeks to travel from Philadelphia to New York seems almost ludicrous. However, in Franklin's time, a letter might take that long to make the 109-mile trip between the two cities. Since roads were neither well kept nor well marked, the most reliable postal routes in the colonies were by sea. Through the combined efforts of Franklin and Folger in publicizing the importance of the Gulf Stream in commerce, the length of time for mail service between major cities was cut in half. Right or wrong, many textbooks and historical references recognize the Franklin-Folger map as the first depiction of the Gulf Stream.

In the minds of many people, Benjamin Franklin is also given credit for having "discovered" the Gulf Stream. While obviously the historical record does not support this view, his contribution to understanding the nature of this current makes a strong case for linking his name with it. In addition to his chart and tireless temperature measurements of the Gulf Stream, Franklin proposed a hypothesis regarding its formation. Like some of his predecessors, Franklin believed that the strong flow of the Gulf Stream was due to the influence of the tropical trade winds. He argued that the wind-driven accumulation of water on the northeastern coast of South America ran downhill in an intense, narrow band into the Gulf of Mexico and then continued on toward higher latitudes. Two interesting points arise from this discussion, which probably can be directly attributed to Franklin's strong persona: First, the Gulf Stream derived its name because Franklin incorrectly believed it originated in the Gulf of Mexico. Second, Franklin's downhill-running flow would remain the dominant idea for the Gulf Stream's strength well into the twentieth century.

In the period between Franklin's first printing of his Gulf Stream chart and its final version, William de Brahm published in 1772 a chart that was more detailed than Franklin's, but it received little recognition. Specifically, his chart shows the Gulf Stream as a component of the overall circulation in the North Atlantic; in contrast, Franklin's map is devoid of the "bigger picture." De Brahm believed that the water pushed into the Gulf of Mexico by the trade winds was compressed by land to form a powerful current that flows through the Straits of Florida and then turns north on its long journey to Nova Scotia. He described this current as joining with the flow from the north before heading across the Atlantic toward the Azores and, finally, south along the African coast to the region of the trade winds. His work is only the second explicit thesis to place the Gulf Stream in the context of the North Atlantic gyre, coming more than a century after Vos first proposed the interlocking nature of the Atlantic's currents. But from a geographical standpoint, de Brahm erred in placing the Gulf Stream too far north along the coast before it turns seaward. This may have had the unfortunate result of placing European ships in the path of the flow that they were attempting to avoid on their passage to America. Whereas de Brahm's chart did not meet muster as an accurate tool for navigating the Gulf Stream, Franklin's chart did succeed, probably contributing to its popularity and high place in nautical history.

Though the gross anatomy of the ocean circulation was developing into a clear picture during the eighteenth century, details regarding specific spatial characteristics of currents, including their boundaries and variability, were still missing from charts and books. The problem was not a lack of human effort or interest but a technological one: determining an exact position on a featureless sea. Once out of sight of land, captains had very few references to pinpoint their location. While European mariners were ascertaining their latitude as early as the fifteenth century, longitude remained the elusive variable. During the sixteenth and seventeenth centuries, the problem of estimating longitude was still unresolved and had become so pressing, because of its navigational and thus economic importance, that European governments were offering large sums of money for its solution. But the prize proved elusive, even to the most ingenious inventors.

The key to determining longitude at sea is having an accurate, reliable clock on board, because keeping track of time is at the core of the measurement of longitude. Up until the middle of the eighteenth century, shipboard clocks were highly inaccurate and too delicate to survive the rigors of an extended sea voyage. Storms, the ship's motion, and the corrosive salt environment took their toll on these fragile clocks. In 1728, John Harrison, a Yorkshire cabinetmaker, developed a new timepiece called a chronometer, which proved to be extremely precise and impervious to the pitch and roll of a vessel. This prototype would ultimately evolve into a more sophisticated version that on a sea trial in 1761, for the first time in history, produced an accurate determination of longitude.

In the late eighteenth century, the widespread use of chronometers aboard ships would lead to an unprecedented period of open-ocean data gathering. In particular, the British admiralty office took the lead in collecting current and weather observations. The downside was that the amount of data being produced was outstripping the office's ability to synthesize it all for the maritime world. This task would fall to James Rennell, who would devote the final four decades of his life to collecting, reducing, and analyzing all the meteorological and current observations made at sea. He was relentless in this task: a monumental undertaking that would culminate in the 1832 publication of a voluminous book and charts of currents in the Atlantic that included the general pattern and variability of these currents, the direction of the prevailing winds in relation to current flow, dates and places of observation, specific depths and temperatures of the sea, and tracks of vessels making important scientific observations. His work would significantly advance the descriptions provided by Vos (1663) and de Brahm (1772) about the interconnections of currents within oceanic gyres and would be recognized for years as a landmark in oceanographic research. In this seminal work, Rennell would devote more than a hundred pages to the Gulf Stream, essentially adopting Franklin's view of it as a flow of water moving downhill from an elevated Gulf of Mexico and Caribbean Sea. In addition, while not specifically addressing the issue of Gulf Stream meanders, he was quick to point out the fact, still relevant today, that the position of the Gulf Stream east of Cape Hatteras is at best imperfectly known.

Recognizing the importance of the oceans with regard to its economic interests, the United States government authorized in 1836 the outfitting of a naval expedition for the specific purpose of surveying parts of the Atlantic, Pacific, and southern oceans. Lieutenant Charles Wilkes was placed in charge of this expedition, which, due to poor planning and bureaucratic wrangling, got off to a bad start. But his leadership and tenacity would prevail, resulting in an exhaustive account of the expedition's findings, which he published in 1845. Ironically, in spite of the wealth of prior evidence championing the concept of wind-driven circulation, Wilkes expressed disdain for theories of the role of the wind in the flow of the Gulf Stream. He viewed the trade winds as too weak to pile up any significant amount of water in the west and to generate such a powerful current as the Gulf Stream. Wilkes went even further in downplaying the impact of atmospheric circulation on the ocean when he categorically rejected the possibility that winds could cause permanent currents anywhere in the ocean. Like many before him, after disowning one theory, Wilkes felt free to promote his own view of ocean circulation. He would use the "hill of water" argument, but he would place the higher sea surface in the center of the ocean gyre. He contended that this elevated sea surface would lead to a downhill flow outward from the Sargasso Sea, and this water spiraling out would account for the circular nature of the gyre. While Wilkes's proposal came closer to the truth than that of his predecessors, his work did not receive wide recognition among his peers, in part because his theoretical explanations for the height of the gyre were weak and unsubstantiated. But Wilkes's fall into relative obscurity was also an occasion of science taking a back seat to hype. The dominating personality of Lieutenant Matthew F. Maury was soon to come on the scene and cast a shadow over Wilkes's accomplishments.

The legacy of Matthew F. Maury is weighted heavily with hyperbole, as he is commonly referred to as the "father of modern oceanography," "scientist of the sea," and "pathfinder of the seas." Even present-day oceanography textbooks consider his work during the nineteenth century to be a landmark contribution to the burgeoning field of oceanography. But others have taken a more circumspect view of Maury and his writings. In particular, a comprehensive paper on the history of ocean circulation by R. G. Peterson and his colleagues in 1996 was quite critical of Maury, claiming that his scientific insights were lacking in depth and often based upon biblical scripture. To which side does the scale tip in judging the merits of this man?

In 1842, Maury rose to the position of superintendent of the U.S. Navy's Depot of Charts and Instruments, where he had unlimited access to the huge and neglected treasure trove of ships' logs from around the world. By 1847, he had assembled this information into coherent wind and current charts, which mariners routinely utilized as navigational aids. In particular, Maury's sailing directions proved to be of considerable value in shortening the sailing time of merchant vessels and thus decreasing their operational costs. In one striking example, he reduced the passage from the eastern coast of North America to Rio de Janeiro by ten days. His fame spread worldwide, and Maury began distributing these charts to mariners in exchange for observational logs of their own voyages. His willingness to dispense his maps stands in sharp contrast to the secretive nature of the Old World traders. But Maury was not a scientist; he was a compiler and was keenly interested in the promotion of maritime commerce, like his predecessor, Benjamin Franklin.

However, Maury was also interested in self-promotion. Realizing that copyright protection did not apply to many of his earlier writings and charts, he published his culminating work *Physical Geography of the Sea* in 1855, assuring him fame and wealth. Maury's lively style of writing, his appealing use of metaphors, such as referring to the Gulf Stream as a river in the ocean, and his vivid imagination all contributed to making his book a commercial success with the public. But this combination of qualities often led him into unfounded generalizations or contradictory speculations about ocean circulation that were soundly rejected by the scientific community. For instance, in discussing the source of the Gulf Stream, he follows in the footsteps of Wilkes, dismissing any influence of winds on this current. As an alternative view, he proposes that temperature and salinity differences in the oceans result in spatial gradients in water density, which initiate a series of vertical and horizontal currents working together to restore equilibrium. So strong was Maury's conviction of the importance of density on flow that he declared it as a *rule* that all currents, whether they are surface or subsurface, owe their origin to this property. His peers bristled at his dogmatic approach to scientific investigation, and they roundly criticized him for lacking the objectivity and discipline necessary for science. In the final analysis, Maury's contribution to ocean science brought widespread popular attention to the field of oceanography, and his theories, though discredited, were beneficial in generating debate among other researchers, who would take the next step in understanding ocean circulation.

The late nineteenth and early twentieth centuries would mark a milestone within the scientific community in arriving at a consensus about the dynamics of ocean circulation. The fundamental idea of wind-driven ocean currents, which had fallen in and out of favor over the preceding 150 years, came back into vogue but now grounded on more quantitative analysis and scientific reasoning. Heretofore, oceanographers had only addressed, through rudimentary observations and qualitative arguments, questions concerning the effectiveness of winds in generating surface currents. A new branch of study, fluid dynamics, which attempted to describe mathematically how fluids behave, would prove invaluable to oceanographers in their quest to understand ocean circulation. But it would take a simple observation to define the connection between the fields of descriptive oceanography and fluid dynamics.

During the voyage (1893–96) of the research ship *Fram* in the Arctic Ocean, the Norwegian explorer Fridjof Nansen observed that ice floes moved at an angle of twenty to forty degrees to the *right* of the direction of the wind blowing across the ocean surface. He was puzzled why the surface water moved in a direction different from the wind. A Swedish physicist, V. Walfrid Ekman (1874–1954), would provide the answer with a fluid model based upon the Coriolis force. Gaspard Gustave de Coriolis (1792–1843), a French mathematician, whose 1835 analysis of relative motion over a surface led to this important finding, is acknowledged by historians for his contribution to the physics of motion, forever linking his name with this phenomenon. Unfortunately, other researchers of Coriolis's time did not immediately recognize the applicability of this principle to the fluid motion of air and water on a rotating sphere like the Earth. Simply, the Coriolis force is the tendency for a moving object, be it an ocean current, wind, or artillery round, to drift sideways from its original path because of the Earth's rotation. In the northern hemisphere, the deflection is to the *right* of the course of motion. (The deflection is to the left in the southern hemisphere, but the following discussion will center, for the sake of brevity, on the northern hemisphere.) What is not simple is attaining a conceptual understanding of this phenomenon.

In a syndicated *Calvin and Hobbes* comic strip, Calvin's father prods him to observe that both the inner and outer points on a record make a complete circle in the same amount of time. Since the point on the outer edge has to make a bigger circle in the same amount of time as the one on the inner edge, it must move faster. The result is two points moving at different speeds, but both completing the same number of revolutions in a given time period. Calvin is totally perplexed by his father's logic, causing him to lay awake at night pondering its significance; but the explanation of it essentially lies at the crux of the Coriolis force.

The horizontal deflection of an object moving across this planet is caused by the Earth's surface rotating eastward at a greater speed near the equator than near the poles (because, like the point on the outer edge of the record, a point on the equator transcribes a larger circle per revolution than one nearer the poles). An object traveling directly toward the North Pole veers eastward (to the right of its original path, if we face in the direction of its travel) because it retains the greater eastward rotational speed of the lower latitudes as it passes over the slower rotating earth near the pole. Similarly, an object possessing the slower rotational speed of higher latitudes and traveling toward the equator tends to fall behind or veer to the west (but still to the right of its original path) relative to the more rapidly rotating earth's surface at lower latitudes.

In addition to seeing the applicability of the Coriolis force to circulation, Ekman's model would also rely heavily on the work of the German fluid dynamicist Karl Zoppritz. His theory that only consistent, prevailing winds are important for inducing currents would gain wide acceptance at the end of the nineteenth century. Ekman now had the two main components — the Coriolis force and steady winds — to develop his "recipe" for gyre dynamics. He would argue that the drag from these persistent winds sets the surface water in motion. But due to the Coriolis force, the direction of the surface flow (wind drift) is at an angle of forty-five degrees to the right of the wind. Now, as a result of frictional coupling in the water column, the next deeper layer will be set in motion, similar to the force of sliding a top playing card over the one below it. But this layer moves more slowly than its surface counterpart and at an *angle to the right* of the overlying water. The same is true for the next layer, and the next, and so on, to a depth of approximately 330 feet. How did Ekman arrive at this number? His mathematical formulation shows that friction will ultimately consume the energy imparted by the wind, and no motion will occur below this depth. Thus, each layer slides horizontally over the one beneath it, each successive layer moving slower and with a greater angle of deflection than the layer above it, until a level of no motion is reached. In summary, the Ekman spiral is the change in speed and direction of the water flow due to the combined forces of wind drag, friction, and the Coriolis force. It is not a true spiral, in the sense of water spiraling downward with depth, like a whirlpool, but rather a way of conceptualizing the horizontal water movements in a column of the ocean. What is important is that the net water

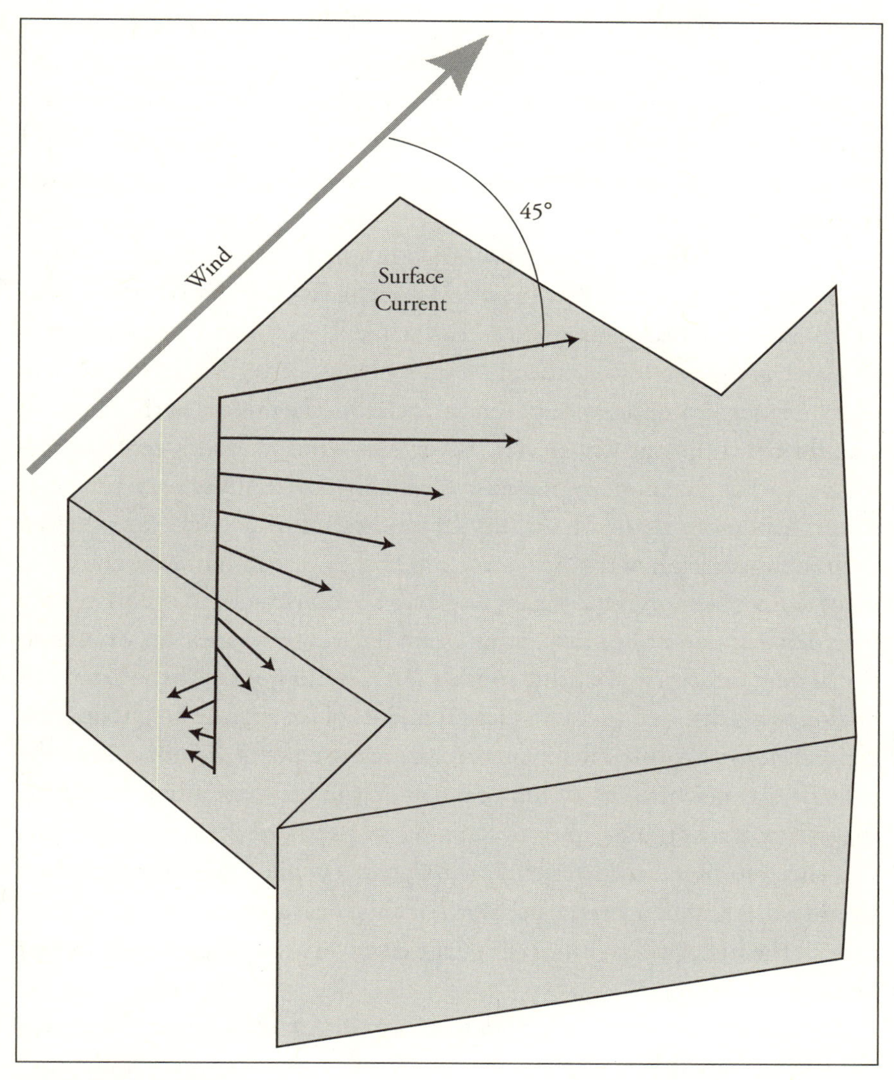

Ekman spiral and transport

motion, known as Ekman transport, over the depth of the spiral is ninety degrees to the right of the wind in the northern hemisphere.

Ekman transport was to become the linchpin in establishing the link between prevailing winds and ocean circulation. If the Ekman model is correct, then in the North Atlantic, there should result a net water movement approximately northward from the northeast trade winds, or perpendicular and to the right of these winds, and approximately southward from the prevailing westerlies in the higher latitudes. In the Atlantic, with its east-west land boundaries, the Ekman transport moves water toward the center of the basin, piling it up higher than along the Atlantic basin's boundaries.

The "hill of water" concept, proposed originally more than three hundred years earlier by André Thevet, is now elevated to its rightful status in the theory of ocean circulation. But while Thevet's idea included a hill of water along the lower latitude flanks of the western Atlantic Ocean, Ekman transport necessitates a central location, the middle of the Sargasso Sea, for this hill. But whose view is correct — the sixteenth-century cosmographer or the modern physicist? Probably no area in science polarizes researchers more than the validity of models: How accurate are they in their representation of the "real" world? Do they oversimplify the complexity of the problem they are meant to clarify? Empiricists and theorists have long debated these questions, often heatedly. In many cases, only observational data can verify scientific models. Thevet did not enjoy the luxury of having at his disposal accurate measurements of sea surface height because the techniques required for these precise measurements did not exist at the time. It was not until 1836 that François Arago drew attention to a leveling survey across the Isthmus of Panama that showed the Gulf of Mexico was three to five feet lower than the Pacific. While subsequent analysis would call into question the accuracy of this measurement, it does not diminish the historical significance of the discovery that they were different heights.

The launching of the oceanographic satellite *TOPEX/Poseidon* in 1992 opened up the distinct possibility of determining ocean topography from space. The satellite's "topography experiment" (hence the name TOPEX) uses a very sensitive radar altimeter to assess the height of the sea surface with unprecedented accuracy. The data from this satellite do indeed show a hill of water in the Sargasso Sea, reaching to an unspectacular height of six feet. However, the gradient of this hill is so low (a change in elevation of approximately one foot every 250 miles of ocean surface) that this relatively "tiny bump" on the ocean surface is not apparent to anyone undertaking a transatlantic voyage. This finding would appear to make Ekman's theory the winner over Thevet's, but as we will soon see, new wrinkles would be added to the evolving picture of ocean circulation.

Considering only the persistent force of the trades and the westerlies, the elevation of this mound should continue to grow as water converges into it as the result of Ekman transport. But the hill does *not* change into a mountain, indicating that equilibrium has been reached. Water responds to the slopes of the ocean surface as it would on land, by running downhill in response to gravity. In other words, the wind forces the water to "flow uphill," but gravity causes it to "flow downhill." As this downhill flow is driven away from the elevated mound, the Coriolis force acts to deflect it back toward the mound. So who wins this oceanographic tug-of-war between gravity and the Coriolis force? Neither. A balance point is reached where the flow of water is neither directed away from the hill nor toward it but instead *around* the hill. When the above two factors balance, the result is a geostrophic (from *geo*, meaning "earth," and *strophio*, "turn") current that moves in a circular path around the mound. So, this discussion, like the water, has come full circle: the observed gyre circulation around the Sargasso Sea is comprised of geostrophic currents in balance between gravity and the Coriolis force. It is important to realize that the ocean currents in a gyre are *indirectly* driven by the wind but are *directly* the result of the geostrophic model of ocean circulation.

The 1930s and 1940s would mark a period of refinement of the geostrophic-flow model because observations did not match the symmetrical, circular flow of water as proposed by the model. In particular, the observed circulation pattern about the gyre is asymmetrical: narrow, deep, and swift western boundary currents, like the Gulf Stream, and wide, shallow, and slow eastern boundary currents, like the Canary Current.

Detailed observations of the Gulf Stream by Columbus Iselin in 1936 suggested that the strong flow of this current was independent of the Straits of Florida through which it flowed. This was in contrast to the claims of other ocean scientists, who argued that the relatively narrow channel caused compression and strengthening of the flow. Iselin pointed out that no suitable explanation had yet been found that would account for the narrowing and intensifying of the Gulf Stream along the western Atlantic. The Gulf Stream was still not willing to give up all of its secrets.

After World War II, Woods Hole Oceanographic Institution was at-

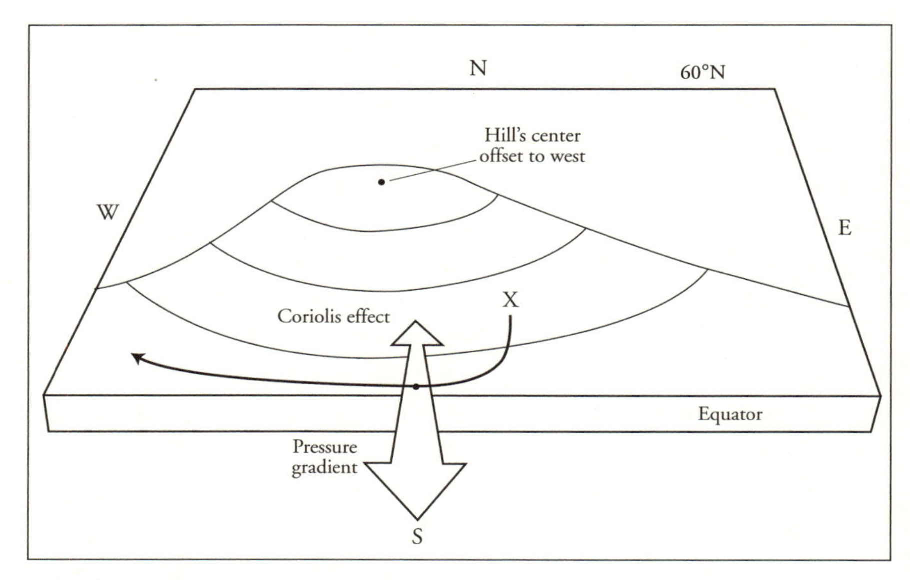

Hill of water and geostrophic flow

tracting many scientists who would specifically study the intricacies of the Gulf Stream flow. The desire by the military for improved understanding of ocean currents in the context of naval operations drove this influx of personnel. One who would bring a sound theoretical footing to ocean circulation was Henry Stommel, who, though originally trained as an astronomer, would gravitate toward oceanographic research.

In an attempt to explain the western intensification of the Gulf Stream, Stommel in 1948 developed a mathematical model of the Atlantic circulation, including a rectangular ocean basin, simplified representations of landmasses, and circulation driven by the trades and westerlies. Though this model, like most, is an oversimplification of the real state of affairs, the simulated gyre that his model yielded was relatively realistic but lacking western intensification. Stommel tweaked his model by adding another factor: the variation of the Coriolis force with latitude. Because the Coriolis force increases with distance from the equator, the eastward-flowing water on the north side of the North Atlantic gyre turns to the south much sooner and more strongly than the westward-flowing water at the equator is turned to the poles. What is the result of this asymmetrical pattern of Coriolis deflection? The apex of the hill is displaced westward from its

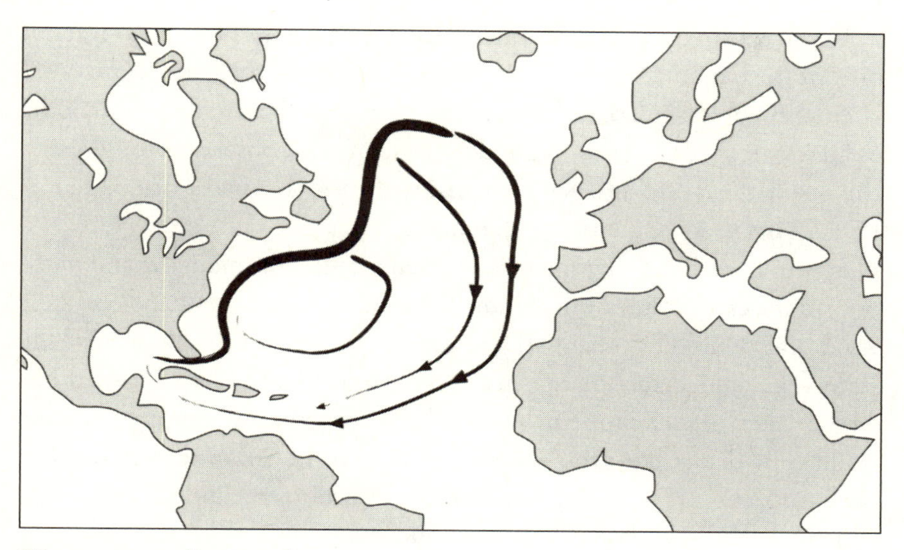

Western intensification of gyre

original location in the center of the Sargasso Sea. Compressed against the western margin, the hill is deformed in its appearance, resulting in a steeper slope on the western edge and a wide, gradual slope on the other side.

Thus, a large amount of water flows south toward the equator across a wide swath of the eastern side of the ocean basin. In contrast, along the western edge of the basin, the water surges poleward in a narrow band. If a constant volume of water is flowing around the hill, then the speed of the current (Gulf Stream) along the western edge has to be greater than the current speed (Canary Current) on the other side of the Atlantic Ocean. This principle is analogous to water flowing within a confined stream channel. As the channel narrows in size, the speed of the stream flow must increase to compensate for the smaller opening through which the same volume of water must be transported. Admittedly, this discussion is a simplification of Stommel's model. A more complete and accurate explanation of western-boundary intensification would include the principle of vorticity, whereby parcels of water not only move in the direction of the current but also spin about a vertical axis. The concept of vorticity involves more complexity than is necessary to dwell on to understand the basic processes at work; therefore, let's acknowledge that vorticity controls in a very direct way the intensification of the Gulf Stream and leave it at that.

But in the final analysis, Stommel's model was an accurate portrayal not only of the dynamics of the Gulf Stream but also of the entire horizontal circulation of the North Atlantic basin. In essence, he had created a model of an ocean gyre that was in good agreement with observations. Though this result appears to be a happy ending to a quest that had spanned centuries, doubts in some quarters still remained.

Stommel realized that his wind-driven circulation model would work only in the case of prevailing winds that generate a persistent and constant wind drag. Fluid-flow scientists commonly refer to this situation as steady state — no appreciable change in the month-to-month or year-to-year observed values of such properties as wind and current flow. Stommel's model could only depict the gross features of the North Atlantic gyre but not the finer details. The changes in current transport, meanders, and rings, which all exhibit variations in space and time, could not be accounted for in the model. But more sophisticated observations in the last half of the twentieth century, which could be meshed with mathematical treatments of the circulation, would yield a better, yet still ongoing, understanding of the whole picture of Gulf Stream dynamics.

The quest for a complete image of ocean circulation was a long and arduous journey, but one fact became inescapable: the atmosphere and the ocean are inextricably linked. Wind drives the ocean's great gyres; the sea, in turn, drives the atmospheric heat engine. The circle remains unbroken. From examining the nature, properties, and dynamics of the Gulf Stream, we now turn to an even more important natural system: the life that thrives in this current, from the smallest organism that floats in the Gulf Stream to the largest that swim its tropical waters.

PART 2 LIFE IN THE GULF STREAM

4 FLOATERS AND DRIFTERS

THE GULF STREAM, in its riverlike fashion, entrains and transports a host of organisms within its boundaries. Most of these creatures, lacking any significant means of locomotion, are carried many miles from their subtropical and tropical origins. Swept along by the rush of the current, they are a diverse group of organisms collectively known as "drifters." While the word drifter often conjures up the image of a vagabond, aimlessly wandering this planet, many of these Gulf Stream drifters are essential providers in the Gulf Stream's complex ecosystem. They provide food, shelter, and on another level, enjoyment to those privileged to peer into the depths of the Gulf Stream. But these drifters often remain elusive and almost invisible. Even to the keenest observer, the Gulf Stream does not readily yield its secrets. Small sizes, transparent bodies, and protective coloration afford many of these organisms a certain degree of anonymity. And to remain anonymous confers a modicum of protection from hungry predators.

Plankton (from *planktos*, meaning "wandering") make up the bulk of the Gulf Stream drifters; they are as diverse as they are inconspicuous. Viewed with the naked eye, plankton appear as tiny specks floating in the water. But collect even a small vial of Gulf Stream water, place it under a microscope, and you will marvel at the incredible variety of bizarre and beautiful shapes: pillboxlike chains, iridescent ovals, and spiked spheres are just a few of the myriad forms that greet a curious visitor to their world. To the uninitiated, these creatures of inner space may seem as bizarre as those romanticized in science fiction movies. Ranging in size from tiny to microscopic, phytoplankton, which are photosynthetic plants, and zooplankton, nonphotosynthetic animals, comprise the major planktonic organisms.

What they lack in size, they make up for in sheer numbers. One hundred thousand cells of the smallest of the phytoplankton, single-celled ultraphytoplankton, would occupy approximately a teaspoon of water. Being small and capable of only limited locomotion necessitates that plankton adapt to their watery environment in order to survive. How do these organisms function?

The key to survival for most phytoplankton is to remain suspended within the Gulf Stream's photic zone, a relatively shallow sunlit layer at about three hundred feet. If they sink below this depth, these vulnerable plants will ultimately perish, deprived of the life-giving light necessary for photosynthesis. An organism will float if it is less dense than seawater and exhibits relatively high frictional resistance to sinking. With regard to the former, numerous planktonic organisms can regulate their density through variations in chemical composition, including producing a tiny droplet of oil that lowers their overall density and increases buoyancy. In addition, many zooplankton become neutrally buoyant by replacing relatively dense chemical elements in their body with lower-density elements. The need to remain buoyant probably accounts for those planktonic organisms that exhibit unique shapes that retard sinking. Elaborate projections or featherlike appendages radiate outward from the body like the spokes on a wheel, effectively increasing the organism's surface area. (Objects that have a large surface area in relation to their volume are effective floaters.) Tropical zooplankton that inhabit the warm waters of the Gulf Stream are generally more ornate with regard to their filamentous projections than their cold-water counterparts. This anatomical difference may reflect an adaptation to the lower viscosity of tropical versus polar seas. Sea-surface temperature strongly controls the viscosity of the water; 40° water is twice as viscous as 95° water. To compensate for the lower viscosity of Gulf Stream water, warm-water zooplankton come equipped with large extensions, effectively increasing their surface area.

Collectively, phytoplankton are the "grasses" of the sea, providing nourishment for the entire oceanic food web. These easily overlooked organisms bind at least 35 billion tons of carbon into carbohydrates each year, accounting for 40 percent of all food made by photosynthetic plants on this planet. Unfortunately, the distribution of phytoplankton pastures is at best spotty throughout the oceans. One could traverse many square miles of the

Diatoms (courtesy of NOAA)

open ocean and never encounter any significant concentrations of phytoplankton. It's as if the ocean were a vast desert, devoid of any significant life. But there is life, and it comes in many forms, sizes, and shapes.

Within this phytoplankton world, diatoms are the most prominent and productive of all of the photosynthetic organisms. As photosynthetic machines, they can effectively convert approximately 55 percent of the sunlight they absorb into usable chemical compounds, one of the highest energy conversion rates known. Under conditions of adequate light and nutrients, the cold Slope Water may experience a diatom spring bloom of enormous proportions, easily visible from space. Since diatoms are capable of biologically producing food at an extraordinary rate, they are the dominant nutritional source for many low-level organisms as well as the foundation of a food pyramid that supports apex predators of the Gulf Stream. As we will see in a later chapter, numerous pelagic species will migrate hundreds of miles tracking the prey that are drawn to the diatom blooms.

While diatoms are more abundant in the temperate zones of the Gulf Stream, dinoflagellates dominate the subtropical and tropical regions. They derive their name, in part, because they possess flagella (small, whiplike structures) for limited movement, giving them some capacity to move vertically into more favorable photosynthetic areas. The dinoflagellates are also known as Pyrrophyta, meaning "fire plants." During one of his voyages to the Gulf Stream, Benjamin Franklin observed, "that it does not sparkle in the night." But little did he know that some species of dinoflagellates cause the water to flash in the blue-green portion of the visible spectrum. These dinoflagellates are bioluminescent (from the Greek *bios*, for "living," and the Latin *lumen*, for "light"); they are capable of producing light from a chemical reaction within specialized cellular structures. Agitation of seawater by something like a fish or a sailboat passing by will stimulate multiple light flashes. Unlike the light produced by incandescent light bulbs, which release some heat, heat is absent from this process, prompting biologists to refer to bioluminescence as "cold light." Large accumulations of dinoflagellates produce "phosphorescent seas," where the crests of waves, including the surf and the bow waves around boats, glow an electric blue. Charles Darwin during his epic voyage aboard the HMS *Beagle* in 1833 marveled at this "living life" display: "There was a fresh breeze and every part of the surface, which during the day is seen as foam, now glowed with a pale light. The vessel drove before her bows two billows of liquid phosphorus, and in her wake she was followed by a milky train. As far as the eye reached, the crest of every wave was bright, and the sky above the horizon, from the reflected glare of these livid flames, was not so utterly obscure as over the vault of the heavens."

Though the light show that dinoflagellates put on can be quite captivating to any seafarer, it may also serve as a survival mechanism. In the Gulf Stream world, where there are no trees or bushes to hide from probing eyes, vulnerable prey must have a defense against predators. When a grazing zooplankton disturbs the dinoflagellates, they emit a flash of light, lasting a few tenths of a second. Like a moth drawn to a flame, a predatory fish is attracted to this flash and feeds upon the larger zooplankton instead of the minuscule dinoflagellates. With their enemy removed, the dinoflagellates flourish, at least for the time being. But the marine environment hosts any number and type of predators, all hungry to feast upon the helpless dinoflagellates.

Copepods, which are microscopic, shrimplike organisms of the subphylum Crustacea, which includes shrimp, crab, and lobsters, graze voraciously upon the dinoflagellates. Copepods, like their bigger cousins, have a hard exoskeleton and a segmented body. They predominate throughout the ocean, outnumbering any other kind of zooplankton. Biologically, they belong to the genus *Calanus*, named after the East Indian philosopher who became an adviser to Alexander the Great and an intermediary for him in his dealings with the Indian sages. Copepods form the link between the phytoplankton that they consume and still other animals that eat these herbivore consumers. In the marine environment, the distinction between "Is that my meal?" and "Am I your meal?" becomes blurred to both prey and predator alike. Larger cousins to the copepods are euphasids, macroscopic zooplankton, which resemble small shrimp and attain a length of about two inches, quite large for the planktonic world.

While size is a common discriminating characteristic in categorizing zooplankton, their tolerance to changes in environmental conditions, like water temperature, is a key factor in determining their geographical range throughout the Gulf Stream and within the current's rings. The type and distribution of plankton in these rings reflect the ecology of their area of formation and the profound biological transformation that occurs with the decay of a ring. The ecology of the cold-water life in the Slope Water is markedly different from the warm-water life in the Sargasso Sea. The Sargasso Sea (and a warm-core ring) is home to a wide variety of tropical and subtropical planktonic species, but the size of the population is small, and the organisms tend to be similar in size. In contrast, the Slope Water (and a cold-core ring) supports a limited number of subarctic species but with a higher population density.

A young ring sustains its initial biological characteristics, but as it decays, the physical environment that supports planktonic life inside it begins to disappear. Studies of the biological changes with time in these rings have shown these transformations are quite complex. A cold-water ring that enters the Sargasso Sea will often contain small, shrimplike crustaceans that are ideally suited to the cold-water environment. If the ring remains isolated from the Gulf Stream, these organisms ultimately disappear from the ring as stirring of the sea by the wind and heating of water by the sun changes the core temperature of the ring, making it unsuitable for habitation. The time required for this extinction varies with the species, ranging from a few months to as long as a year and a half. The decay of the ring occurs most rapidly in the surface layer due to atmospheric mixing and heating. In response to this initial change, these adaptive zooplankton vertically migrate to the deeper portion of the ring that still exhibits its original thermal characteristics. But unfortunately for the zooplankton, one method of demise is traded for another. Their downward movement means that they become separated from their food source, the phytoplankton. Deprived of nourishment, the shrimplike organisms undergo physiological and biochemical deterioration, resulting in slow starvation. Warm-water species of the Sargasso Sea soon take advantage of the bad luck of the cold-water zooplankton. With the decaying ring offering a more suitable temperature for their needs, these opportunistic zooplankton readily move into their new habitat. In fact, some of the warm-water species are more abundant in these rings than they are elsewhere in the Sargasso Sea. One study showed that over a four-month period the abundance of a small snail in a ring had increased by a factor of three hundred. While for humans home may be where the heart is, to some zooplankton of the Sargasso Sea home may be where warm-core rings provide optimum living conditions.

In addition to taking up residence in Gulf Stream rings, some zooplankton are part of a curious underwater feature. Each evening as the sun sets over the Gulf Stream and the nearby Sargasso Sea, millions of zooplankton, mostly euphasids and copepods, slowly ascend from the ocean depths to feast upon the microscopic plants residing in the surface waters.

Feeding under the cover of darkness, they avoid detection by their predators. Unmolested, these zooplankton feed continuously throughout the night, but with the approach of daylight, they reverse course, sinking ever so slowly to reside in the darkness during the day. Biologists have known about this vampirelike behavior since the late 1800s because their sampling nets came back fuller at night than during the day. But the extent of the vertical movement wasn't discovered until World War II when the U.S. Navy was testing sonar equipment to detect enemy submarines. On many of these recordings, a puzzling sound-reflecting surface appeared that was too shallow to be the bottom. Initially, naval personnel referred to this strong reflecting surface as a "false bottom," but upon further analysis, they changed the name to the "deep scattering layer" — a densely packed layer of organisms that looks like a nearly solid surface hanging in midwater.

These tiny animals undergo an extensive vertical migration on a daily basis, ascending more than fifteen hundred feet each evening, then returning in the morning. As they ascend and descend throughout the water column, they experience a total pressure change of approximately seven hundred pounds per square inch of their surface. How do these relatively fragile creatures survive the crushing pressure? We must look again at the anatomy and structure of these unique organisms. Their bodies are essentially devoid of gases, like oxygen and carbon dioxide, which are compressible into a smaller volume when pressure is applied to them. Because the body cavities of many zooplankton contain only water — a fluid that resists mechanical compression — they are unaffected by this high pressure. In addition to the change in pressure, this excursion takes a lot of energy because it requires traveling literally tens of thousands of body lengths every day. Biologist Deborah Steinberg, who has systemically studied these organisms, compares this daily migration to a person walking twenty-five miles to get back and forth to his next meal.

Steinberg's research interests extend beyond studying the feeding habits of these animals and include their possible role in climate. How can these tiny creatures have an impact on potentially a global scale? What is the link between biological processes and atmospheric events? She speculates these migrating animals consume vast quantities of plant material, which affects the carbon dioxide balance between the ocean and the atmosphere. Carbon dioxide is one of the Earth's main greenhouse gases, known to maintain higher global temperatures. Since the plants that zooplankton eat are rich in carbohydrates, which have been synthesized from carbon dioxide and water by using the energy from sunlight, they are potentially moving a tremendous amount of carbon from the surface layer to the ocean depths. This unique interaction of microscopic plants and animals in the Gulf Stream may be playing a small, albeit important, role in regulating our planet's temperature.

Even to the most casual reader, it may be hard to accept that very little was known about the nature of planktonic life in the ocean until more than four hundred years after Columbus's voyages. These tiny critters did not divulge their secrets, in part, because their minuscule size and elusiveness made them a collector's nightmare. But the pioneering expedition of the German research vessel *Meteor*, which crisscrossed the Atlantic during its two-year voyage in the early part of the twentieth century, would help to alleviate this dearth of knowledge. It was the first large-scale, systematic study of plankton and set the standard for future planktonic studies; many of the tools and techniques are still in use today. The *Meteor* biologists employed the use of plankton nets, hauled slowly for a prescribed distance behind the ship, to sample the planktonic life. These cone-shaped, fine-meshed nets trapped the organisms, which were ultimately flushed from the nets to assess their abundance and type. Haul after haul yielded new forms of life, mesmerizing even the most jaded crew member, but the on-board scientists were particularly enamored with the quantitative data that these samples were now providing. For the first time, it became possible to estimate the all-important plankton biomass, the total mass of these organisms in a given area or volume of habitat. Though today's seagoing biologists would not leave port without plankton nets, they and even their predecessors soon came to realize that these nets have a number of disadvantages: very small plankton can slip through the mesh, the nets may crush the delicate plankton during retrieval, and the time-consuming routine of towing the nets limits their sampling capability. The latter problem is particularly troubling since it constrains scientists to obtain information from only a relatively small area. The question is, how representative are these measurements when they are extrapolated to a larger study area? The launching of the satellite SeaStar in 1997 opened a whole new window of opportunity for phytoplankton assessment. A color scanner, the

SeaWiFS (Sea-viewing Wide Field-of-view Sensor), was on board to monitor chlorophyll levels. Since chlorophyll is an essential pigment found in photosynthetic plants, the measurement of chlorophyll correlates directly to the phytoplankton biomass—high chlorophyll concentrations mean high plant biomass.

Not surprisingly, these images have shown there is very little phytoplankton biomass throughout most of the Gulf Stream. Lacking the nutrients needed for photosynthesis, the phytoplankton biomass in the Gulf Stream is approximately twenty-five times less than that in the neighboring nutrient-enriched Slope Water. On a SeaWiFS image, the pea-soup green of the Slope Water stands in sharp contrast to the blue of the Gulf Stream.

One intrepid entrepreneur, Michael Markels, is convinced the low plant biomass of the Gulf Stream could be significantly increased by ocean farming techniques. Markels believes the phytoplankton biomass could swell by a factor of about one thousand to a billion tons per year through the addition of nutrients, such as phosphorus and in particular iron, to the water. He proposes fertilizing the water in much the same way a farmer fertilizes a field. In turn, he estimates, that could increase the fish catch by a factor of four hundred, from 125,000 tons to 50 million tons per year. Also, since all of the ocean farming and fishing would occur within the U.S. Exclusive Economic Zone, all the economic benefits from the increased fish production would accrue to the American fishing industry. Sound too good to be true? Possibly. At least, that's the consensus of some scientists. Markels's enthusiasm for his project rests upon the iron hypothesis, which goes back as early as the 1930s, when the potential role of iron in phytoplankton photosynthesis was first touted. In principle, sufficient iron concentrations in the water are needed for the complete assimilation of nitrates and phosphates by the phytoplankton. (We will see later in this chapter that iron from the wind-blown soils of Africa may contribute to the vitality of plant growth in the Sargasso Sea.)

In the early 1990s, two field projects (IronEx) tested this idea by adding iron to approximately a twenty-five-square-mile patch of equatorial Pacific Ocean. Both tests yielded significant phytoplankton blooms, but even these positive results raised more doubts about the applicability of the experiments to the Gulf Stream. Sallie Chisholm of the Massachusetts Institute of Technology, who participated in both of the above experiments, argues it is not that simple to extrapolate the results from one area to a totally different region, characterized by distinctive physical and chemical properties. While the test site in the equatorial Pacific is relatively rich in essential plant nutrients, only lacking the iron catalyst, the Gulf Stream is nutrient poor. No amount of iron can compensate for this deficiency. In addition, the IronEx projects were on a relatively small scale compared to the proposed large project for the Gulf Stream. Dissidents huffed that it is equivalent to comparing an urban garden plot to a thousand-acre Iowa cornfield.

While proponents of ocean farming claim the massive addition of nutrients to the ocean would be environmentally benign, fertilization does change the composition of the phytoplankton community, which, in turn, would have repercussions up the food chain. Even if large-scale ocean fertilization proved to be successful in stimulating massive phytoplankton blooms, there are major hitches to ocean farming the Gulf Stream: it moves, meanders, and spins off eddies and rings. As a result, where the nutrients are added would not necessarily coincide with where the fish are caught. If, for example, waters were fertilized off Key West, this would yield improved fishing off north Florida and Georgia, assuming a Gulf Stream flow of four knots and a delay time of four days from fertilization to yield. Since there are no fences on the Gulf Stream, how would it be possible to reap the benefits from increased fish production? And who would profit from increased fish yield? The region of concern is a commons — an area not owned by specific individuals. Without definitive and documented private property rights, what would be the economic incentive to invest in ocean farming if the yield were reaped by other individuals? A common public perception is that the catching of fish is synonymous with "harvesting the sea." The use of the word "harvest" is misleading because while farmers have a financial investment in their land, plant seed, cultivate crops, and control pests, fishermen take the fish for free since the marine "pasture" is open to all. The analogy between farmers and fishermen disintegrates even further because the farmer must ultimately replant his crops once they are harvested. The fisherman is involved in the taking of the resource, not in its replenishment. In the absence of property rights, each rational fisherman seeks to maximize his individual gain: catch as many fish as possible before someone else does. The seminal work "The Tragedy of the Commons" by Garrett Hardin is the foundation for much of the current thinking with regard to exploitation of biological marine resources. Though Hardin chose unregulated cattle grazing as his example, he argues each individual is locked into a system that compels him to increase his gain (catch) without limit, but in an environment (Gulf Stream) that is limited.

In addition to the aforementioned microscopic phytoplankton and zooplankton are animals found on the other end of the size spectrum, classified as megaplankton. Foremost in this group are the cnidarians. All of these members have soft, jellylike bodies, which are more than 95 percent water, and have long, flowing tentacles. Cnidarians are divided into two basic subgroups: the hydrozoans and scyphozoans (jellyfish). The largest member of the hydrozoans (from *hydro*, "water," plus *zoa*, "animal") is one of the most common and intriguing Gulf Stream species, the Portuguese man-of-war. This drifter is not a single animal but a complex form of life, made up of a colony of several different types of polyps. Each has a special function, such as digestion or reproduction, and they work together to ensure their mutual survival as floating houses of the sea.

Even the novice seafarer can recognize the Portuguese man-of-war's translucent, iridescent, gas-filled float that may be three to twelve inches long and extend six inches above the water surface. Portuguese sailors were the first to give this species its moniker, man-of-war, because to them its float appeared similar to the sails of the massive, three-masted man-of-war ships. Like its namesake, the Portuguese man-of-war uses its float as a sail, gathering the wind and skimming across the ocean surface. Residing in the offshore waters of warm seas, the Portuguese men-of-war drift northward during the summer months. The combined influence of the Gulf Stream and southerly winds may push them as far north as Cape Cod. Much like a storm-tossed sailboat furling its sails to better weather heavy seas and strong winds, the Portuguese man-of-war can deflate its gas-filled float in order to sink just beneath the water surface, where it will stay out of harm's way if conditions warrant it during the drift to higher latitudes.

On this northward journey, these organisms may cluster — some even say "swarm" — into large groups numbering in the thousands. Gathering in drift lines caused by the current, they may form a mass of organisms that stretches for miles across the open sea. And like a fleet of warships ready to enter into battle, they make a formidable armada. Each organism

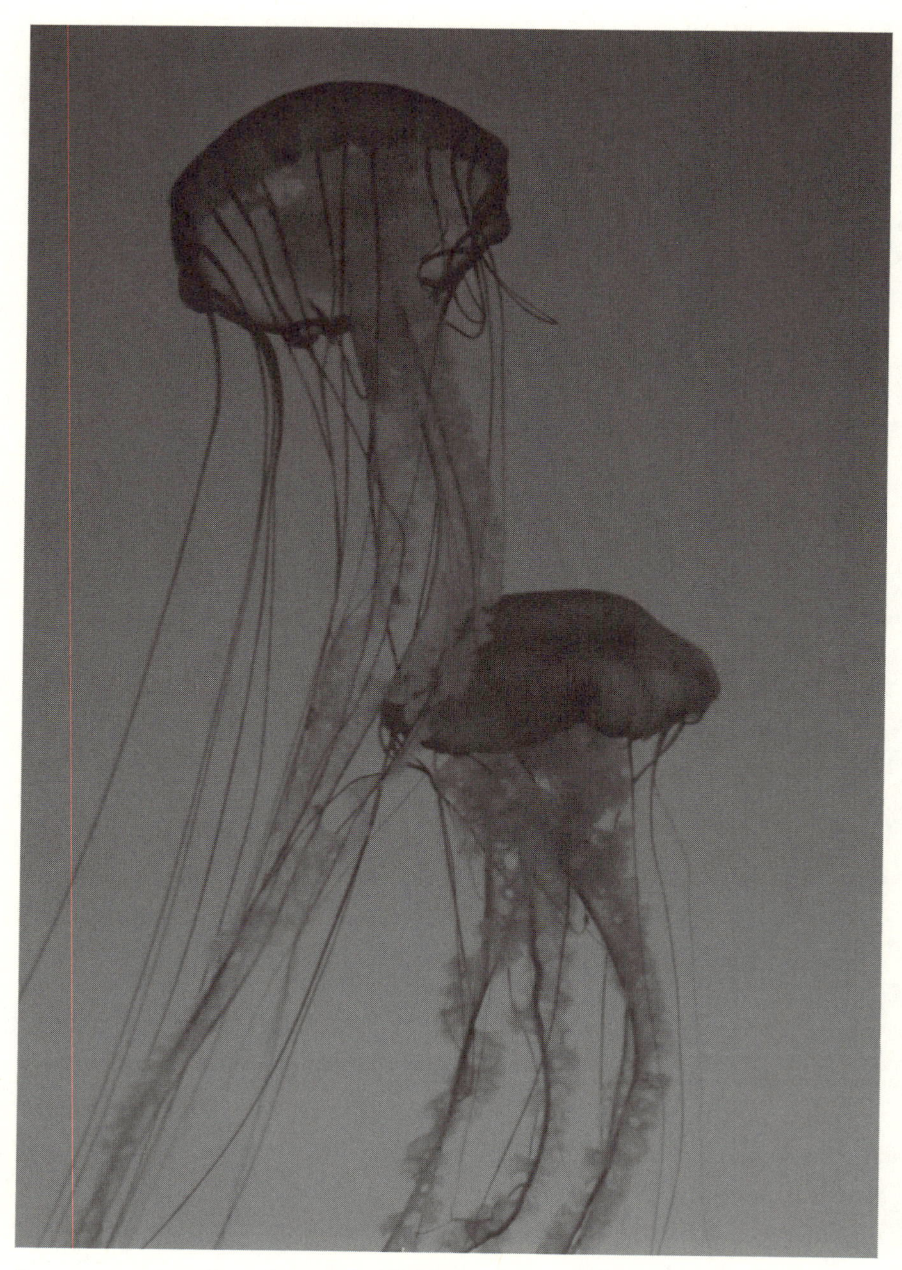

Portuguese man-of-war (copyright Elzbieta Szpak/Shutterstock.com)

comes armed with a great number of tiny stinging cells, called nematocysts, which reside in a maze of tentacles that hang fifty feet or more from the float. Each cell contains a coiled hollow tube tipped with a barb. When the cell comes in contact with a prey or experiences any external pressure, the nematocysts discharge, much like a miniature harpoon, into the prey. The sting contains a powerful neurotoxin, almost as strong as cobra venom, which results in paralysis of the prey. Muscles in each tentacle contract and drag the unfortunate victim into range of the digestive polyp that secretes a wide range of enzymes and breaks down the prey's protein, fats, and carbohydrates. As the Portuguese man-of-war drifts within the Gulf Stream, the long tentacles "fish" continuously through the water column. It takes no prisoners; it will eat basically anything: small fish, crustaceans, and other members of the surface plankton that become ensnared in its entangling threads.

Although the stinging cells of the Portuguese man-of-war are capable of killing a large mackerel, a small fish called the man-of-war lives with impunity among the deadly nest of tentacles. A symbiotic relationship characterizes the intimate living arrangement between these two different species, and this unique state of affairs poses two questions: what does our intrepid intruder have to offer its companion? How is the man-of-war fish able to avoid being killed by its potent host? The answer to the first question may depend on the two parties finding this relationship mutually beneficial. Most likely this small boarder receives protection from larger predators by taking up residence among the tentacles, but it may earn its keep by enticing unsuspecting prey into these same tentacles. Arriving at a conclusive answer to the second question is a bit more elusive. The fish feeds on some of the tentacles, and it is possible that ingestion of some of the venom helps it to build up immunity against a massive discharge of these stinging cells. In laboratory experiments, the fish survived high-dose injections of venom, while a different fish of similar size succumbed to a dose only a tenth of that received by the man-of-war fish.

A close relative of the Portuguese man-of-war is the small, blue-colored, and colorfully named by-the-wind sailor. But instead of the gas-filled float found in its cousin, it has a flat disk with a thin, semicircular fin set diagonally along the disk that acts like a sail, from which the species takes its name — *Velella velella* (from *velum*, "sail"). And quite a good sailor it

By-the-wind sailor (author's photo)

is, ably tacking with the wind. Interestingly, the orientation of the sail along the disk determines which way *Velella velella* will travel. If the sail runs northwest to southeast along the disk, it will drift to the left of the wind direction. A drift to the right of the wind occurs with a sail oriented southwest to northeast. At the mercy of Atlantic storms, these organisms wash ashore in large numbers, but these strandings tend to be composed of only one group — those that either drifted left or drifted right of the wind. Though the by-the-wind sailor does have stinging cells on its short tentacles to capture planktonic organisms, these cells have low-power stinging "batteries" that are essentially harmless to humans. In contrast, its fellow traveler, the Portuguese man-of-war, even though appearing lifeless on a strand of beach, will still possess a potent arsenal of stinging cells. The very painful sting can cause a series of red welts on the skin, and the venom can result in serious health effects, including fever, shock, and interference with heart and lung functions. Like the pirates and scoundrels who feared its warship namesake, today's more benign sea adventurers should assiduously avoid the Portuguese man-of-war.

True jellyfish, or scyphozoans, have a bell-shaped float from which hangs a fringe of tentacles. This anatomical configuration is the characteristic medusan body form, named after Medusa of Greek mythology. She had once been a beautiful woman — her hair being her chief glory — but upon challenging the goddess Athena, her ringlets were changed into hissing serpents. Deprived of her charms, Medusa became a hideous monster, so frightening that no human could behold her without being turned into stone. In contrast, present-day observers of the drifting medusa-shaped jellyfish experience a sense of wonderment at these beautiful creatures. Serenely riding the Gulf Stream, jellyfish control their vertical movements by a series of muscle contractions. Water enters the cavity under the bell, and the muscles that encircle the bell force the water out. This propulsion mechanism allows the animal to jet ahead in short spurts. The capacity for locomotion is important because jellyfish feed by swimming to the surface and then sinking, like a deflated balloon, through the rich surface layers. But its fragile beauty and subtle motions can be deceiving, because its tentacles have serpentlike qualities — able to inflict a venomous "bite." The potency of the venom found within the stinging cells is proportional to the size of the jellyfish. Most jellyfish are less than two feet in diameter, but some may reach over five feet in diameter, trailing tentacles over fifty feet in length. While human contact with the medusa does not turn an unfortunate individual into stone, a once-pleasant observing experience may indeed become a painful encounter.

The nature of the jellylike organisms that float within the Gulf Stream can range from the dangerous to the benign. The Gulf Stream doesn't discriminate with regard to its passengers; it is simply a conveyor of these drifting forms of life. Vacationers, enticed by the warm waters of the Outer Banks, have sometimes found an ocean with the consistency of Jell-O and piles of small, jellied organisms that have spun off from the Gulf Stream and washed ashore. Beachgoers want to know what they are and whether they pose any danger, like their counterparts with tentacles. The guesses include shark eggs, fish roe, and even juvenile jellyfish. In fact, they are called salps, animals composed of a gelatinous tube with an opening at both ends. Sucking water in one end and propelling it out the other serves as both a highly efficient means of propulsion and a means of filter feeding, because the intake water must pass over a membrane before it exits the organism. The membrane vacuums the water clean of any food material. Though gen-

Sargassum weed (courtesy of NOAA)

erally no more than three inches in length, these organisms can form jelly chains that are more than five feet in length. In the clear waters of the Gulf Stream, where shelter and protection are at a premium, small fish may be seen clustering beneath these aggregations of salps. If conditions are right, they can form huge blooms; the population commonly increasing more than a thousand times from one year to the next. A common breeding ground is the Georgia Bight, a large expanse of water between the Georgia coast and the Gulf Stream. Entrained into the current, these open-water species are only visible to beachgoers when an onshore wind or eddy brings them closer to shore. The sticky aggregation of salps can be a nuisance to bathers, but these translucent organisms are harmless, neither stinging nor biting. Though the Portuguese man-of-war, by-the-wind sailor, jellyfish, and salps are all gelatinous in nature, appearing soft, mushy, and without the dental structure of the more feared residents of the ocean, the sight of tentacles should be a "red flag" for any traveler on the Gulf Stream.

The signature drifter of the North Atlantic gyre is the plant known as sargassum weed (actually a brown algae) that gives the Sargasso Sea its name. Often occurring as vast mats floating in the seemingly endless expanse of the Sargasso Sea, there are millions of tons of this weed adrift in this "eye" of the gyre. Encircled by the North Atlantic gyre and the prevailing winds at these latitudes, the sargassum spreads over more than a million square miles of the tropical and subtropical Atlantic. Because of the interlocking nature of the currents, little of the sargassum ever exits its languid home to become part of the rest of the tempestuous Atlantic. Of the sargassum that does exit the Sargasso Sea, prevailing winds, storms, and currents disperse it throughout the Gulf of Mexico, the Caribbean, and the western Atlantic. Specifically, some of the clumps and pieces of sargassum may drift through the Caribbean basin, pushed along by the Caribbean Current through the Yucatan Straits into the Gulf of Mexico. Once there, the clockwise-flowing Loop Current carries the sargassum northward. Eddies, spinning off the current, bring the offshore waters and their associated load of sargassum closer to shore. Continuing on its journey, some of the material hitches a ride on the Gulf Stream and works its way up the Atlantic coast. Throughout this region, the sargassum frequently aggregates into long, thin rows, called flotsam lines. These linear patches form in response to wind forcing or shearing currents along the frontal boundary of the Gulf Stream. Ultimately, the whirling rings of the Gulf Stream entrain the sargassum, bringing it either closer to shore or back to the Sargasso Sea. This plant is truly one prodigious drifter.

Sargassum gets its name from the pea-sized floats, containing a mixture of mostly oxygen, nitrogen, and carbon dioxide, which provide for its buoyancy. As Portuguese sailors journeyed to the New World, they christened the floating rafts of weed *salgazo* because the tiny floats brought to mind a small grape found in their homeland. Over a period of time, the Portuguese word for gulfweed, *sargaço*, came into favor; later this species of brown algae was classified under the genus name *Sargassum*. Upon encountering great floating masses of sargassum during his first voyage to the New World, Christopher Columbus described it, simply, as "weed" and recorded its abundance in his log on September 17, 1492. While it is almost certain that Columbus was well aware of the existence of seaweed from his early voyages to the Azores, his crew feared the weed would become so thick and tangled as to irretrievably ensnare their ship. This superstition dates back hundreds of years before Columbus's voyage. The Carthaginian admiral Himlico (ca. 380 B.C.), after sailing through the Pillars of Hercules (present-day Strait of Gibraltar), reported: "Many types of seaweed grow in the troughs between the waves, which slow the ship like bushes. Here the beasts of the sea move slowly hither and thither, and great monsters swim languidly among the sluggishly creeping ships." While the fears of Columbus's crew were misplaced — the sargassum is neither thick enough nor structurally built (flimsy branched stems with numerous leaflike blades) to stop a ship — these old myths persisted well into the nineteenth century. Detailing the adventures of Captain Nemo in *Twenty Thousand Leagues under the Sea*, Jules Verne vividly described the perils of the Sargasso Sea: "Such was the region the Nautilus was now visiting, a perfect meadow, a close carpet of seaweed, fucus, and tropical berries, so thick and so compact that the stern of a vessel could hardly tear its way through it. And Captain Nemo, not wishing to entangle his screw in this herbaceous mass, kept some yards beneath the surface of the waves." Paintings during this era depicted sailing vessels being devoured by the sargassum, and the tabloid press promoted the idea that freighters sat becalmed and weed shrouded along with the sailing ships from Roman times, for nothing ever changed in this stagnant sea. To Columbus, sargassum was more than just an oddity, because he considered it a sign that he was close to land. Most of the seaweeds he knew about were nearshore species that were anchored to the bottom. With the hope that land was nearby, he sounded the sea, only to find no bottom in the center of the Atlantic Ocean. He was still more than three weeks from landfall in the New World. But Columbus would learn from his mistake, and on his return voyage, he would refer to the observations, which he had recorded in his log on the location of sargassum, to make an adjustment in the ship's heading.

While Columbus had no formal training in marine botany to aid in the recognition of this "weed," scientists have classified at least six known species of sargassum. But ultimately two distinct species, *Sargassum natans* and *Sargassum fluitans*, proliferated into the millions of tons of weed now adrift in this pelagic environment. The two species are quite similar in appearance, and many mats and weed lines contain both species. As to the origin of these species, growing hundreds of miles from any land, Caribbean storms, millions of years ago, may have uprooted the nearshore plants from their bottom habitat. Buoyed by their gas-filled floats, these detached clumps managed to survive and ultimately evolved into the weed that Columbus encountered on his voyages. Sargassum readily adapted to its new lifestyle because it reproduces asexually through fragmentation; thus every piece of this species may be traced back to a single ancestor. Once a bit is dislodged, it can essentially reproduce over and over again, almost indefinitely. But sargassum does not live or drift forever, or else the sea would indeed become choked with it. As clumps of the seaweed become encrusted over time, its natural buoyancy is overcome by the increased weight. When it sinks below the critical depth of approximately three hundred feet, the hydrostatic pressure will crush the floats, leaving only a slow descent to the cold, perpetually dark bottom of the Sargasso Sea for these organisms.

Paradoxically, the success of sargassum occurs in a nutrient-impoverished region of the Atlantic Ocean. From an oceanographic point of view, the Sargasso Sea is more than a surface phenomenon, but rather extends thousands of feet to the bottom of the Atlantic Ocean. Its most distinguishing physical characteristic is an elevated lens of warm, transparent water sitting over the cold abyss. The strong solar heating and calm conditions imparted by the Bermuda high-pressure center ensure that very little, if any, nutrient mixing occurs between these layers. It's as if a lid had been placed on the cold, nutrient-rich deep water. Thus, the Sargasso Sea supports a thin gruel of phytoplankton organisms. Marine biologist Henry Genthe, who has crossed the Sargasso Sea at least six times, has estimated that during the course of a year, one square yard of this sea produces the phytoplankton equivalent of weekly grass clippings from a one-square-yard patch of lawn.

So where does the nourishment to support vast fields of sargassum come from? The answer might just be found by looking at the trade winds — the consummate conveyor of ships and sailors alike. The arid regions of northern Africa, including the Sahara and the Sahel, supply large quantities of wind-eroded soils, mostly in the form of dust, which are carried by the winds to distant parts of the Atlantic basin. In situ and remote sensing measurements place the amount of this so-called aeolian (from Aeolus, a Greek god who kept the winds bottled up in a cave) material at hundreds of million tons per year. Ground-based studies on the island of Barbados, dating back to the 1960s, were the first to show that this dust can travel across the Atlantic and subsequently fall out over the Caribbean basin. But not all this dust travels the breadth of the Atlantic; significant deposition of this material occurs in the North Atlantic. This material influences the nutrient dynamics of the marine ecosystem by supplying a number of nutritive species, such as nitrates, phosphates, and, maybe most important, iron from the lateritic (weathering of underlying rock) soils of Africa. The amount of nutritive species transported and deposited to the ocean may wax and wane, depending upon the frequency and intensity of Saharan dust outbreaks. Rather than a conveyor-belt-like delivery of dust, these outbreaks are episodic and discrete events. Attempts to quantify the amount and distribution of sargassum in response to variations in nutritive loading are fraught with difficulties and are just beginning to be addressed by the scientific community. It may be that the vitality of the sargassum community can be found as the songwriter Bob Dylan has penned, "blowing in the wind."

Pelagic sargassum supports a diverse assemblage of organisms, including fungi, micro- and macro-epiphytes (plants that grow on other plants), and dozens upon dozens of invertebrates, fish, eels, and turtles. The ability of sargassum to remain buoyant, its seasonal distribution, and its specific geographical location determine the species composition and abundance. This web of life comprises a "floating jungle." The high diversity of organisms, dependent upon this marine jungle, may be the result of a number of factors: protection, shade, structural affinity, feeding opportunity, and use of spawning substrate. With its tangled maze of stems, fronds, and floats, sargassum provides a perfect shelter for drifting larvae and small fish to take up residence and remain secluded from the probing eyes of potential predators.

Of the hundreds of denizens that call sargassum home, two species of fish have evolved particularly unusual shapes and coloration, affording them the advantages of camouflage and stealth among the floating rafts of sargassum. The chief resident of the sargassum is its namesake, the sargassum fish—golden-brown in color, accented with patches of white, perfectly matching the color and pattern of its host. With its unique coloration, ragged appendages that mimic sargassum fronds, a head and body that appear as one, it is hard to tell fish from plant. When you are just three

Pipefish (copyright Annetje/Shutterstock.com)

inches long, sargassum makes a good refuge from bigger nemeses. Using its peculiar armlike pectoral fins and fin rays splayed out like hands, the sargassum fish crawls through its jungle home, grasping the stem of the weed and pulling itself along. Swimming is not an option for this forest dweller. But the sargassum fish are not always passive, nor do they lead a totally sedentary existence. What they lack in stature, they make up for in attitude as aggressive and feisty hunters of the sargassum jungle. These pelagic bantamweights will make a meal of anything that comes along, including their own kind, as long as it can fit in their gaping maw. They entice their prey into striking range by wriggling a fleshy flap of skin that looks like a tiny worm. The motion piques the curiosity of its prey, and when it comes closer, the sargassum fish inhales it with a jetlike suction from its mouth.

A neighbor of the sargassum fish is the pipefish, a close relative to the seahorse. A small fish, its elongated shape and brownish-green coloration make it well-suited for blending in with the sargassum weed. Its slender, coiling body is ideally suited for navigating between and among the thickets of sargassum as it searches for small zooplankton, mysid shrimp, and various invertebrates. Like many of the organisms that inhabit the sargassum, pipefish survive by being omnivores, not limiting themselves to a single food source but making a meal of whatever creature it can find. In contrast to the sargassum fish, the mouth of the pipefish is small, located at the end of a tubelike snout, giving rise to its genus name *Syngnathus*, meaning "jaw together." Though lacking teeth, it is not likely to pass up a meal. This feeding machine can swallow a whole prey and rapidly pass it through its digestive system, ready to eat again.

While most of the sargassum community is generally made up of permanent residents, there are visitors. Each year millions of eels from the rivers of North America and Europe congregate in the Sargasso Sea, which becomes the international meeting place for eels, intent on one purpose: spawning. Exactly where they go in the Sargasso Sea is still a mystery since, paradoxically, no live adult eels have been found there. Most likely, they swim down into the abyss where they mate, lay their eggs, and die. This portion of the life cycle of eels remained unknown for centuries. The philosopher and naturalist Aristotle was probably the first, around 350 B.C., to put forth a hypothesis in his *History of Animals*. He thought eels arose spontaneously from the mud of river bottoms. About 60 A.D., Pliny the Elder wrote that adult eels reproduced by rubbing their skin against rocks, the scraped-off pieces transforming into young eels. Even as late as the eighteenth and nineteenth centuries, eel connoisseurs were puzzled by a major riddle. Why were no baby eels ever found in the freshwater rivers and creeks of Europe and America? In 1922, a Danish oceanographer, Johannes Schmidt, solved the puzzle when he discovered the European eels' spawning ground in a remote section of the Sargasso Sea, northeast of Cuba. The eels had not surrendered this information readily. Schmidt dedicated eighteen years of his life — tirelessly trawling the Atlantic, collecting young eels, and attempting to pinpoint the Sargasso Sea site with the greatest concentration of eel larvae.

Additional scientific sleuthing uncovered that en masse spawning migration to the Sargasso Sea occurs between January and March. Though Columbus developed the eye of a naturalist, observing the flights of birds and the drift of weed, he would have been hard-pressed to see the tiny, transparent, leaf-shaped eel larvae, which hatch and ascend to the surface after the adults have mated. These newly hatched eels, called leptocephali — or, in a more colorful phrase, slime heads — are so different in appearance from adults that they were originally classified as a separate species of fish. Drifting homeward on the Gulf Stream and the other currents of the North Atlantic gyre, American larval eels will reach their destination in a year or more, but it will take two years for their European cousins to finally reach their destination. During this period, their only sources of nourishment are plankton that they filter from the water, but they also undergo metamorphosis to more closely resemble a small eel, elver, or glass eel. Upon detecting the chemical signature of their parents' home river or estuary, they gather in huge numbers at the mouths of these inland waterways to begin their journey upstream. They will return to the Sargasso Sea upon reaching sexual maturity around the age of ten to begin the cycle again.

A reverse trip occurs for visiting sea turtles. While turtles lay their eggs on coastal beaches, posthatchling sea turtles, including loggerhead, green, Kemp's ridley, and hawksbill, will enter into a pelagic stage. During the first year or more of their lives, they are found drifting in the sargassum rafts that gather in the Gulf Stream. In addition to seeking protection within the seaweed, the turtles apparently subsist on it: an analysis of the stomach contents of these hatchlings showed that almost all contained sargassum floats and leafy parts.

Because of the unique assemblage of organisms within the sargassum community and the large presence of juvenile fish, which attract predatory game fish, the South Atlantic Fishery Management Council (SAFMC), the federal agency responsible for protecting pelagic fish and their habitat from North Carolina to Florida, has declared sargassum as "essential fish habitat" under the Magnuson-Stevens Fishery Conservation and Management Act. Guided by this law, the council is charged with minimizing any adverse effects on such habitat. But is it protected? And from whom? Dating back to the 1970s, a small commercial fleet, based in North Carolina, harvested sargassum, albeit in small amounts, from the waters of the Carolinas. This burgeoning enterprise processed the sargassum into nutritive supplements and feed additives. With the concern that this harvesting would escalate and negatively impact the amount of sargassum, SAFMC entered into a six-year struggle, which grew into a bureaucratic nightmare, to shield this resource. The effort culminated in a management plan that became law on November 1, 2003, to curtail the development of a widespread fishery for the seaweed. The Sargassum Management Plan imposes strong limitations: prohibition of harvesting south of the North Carolina/ South Carolina border, limitation of the annual harvest to five thousand pounds, and a ban on harvesting within one hundred miles of the coast. While this was a bold initiative in protecting sargassum throughout much of the U.S. EEZ, the ability to secure protection in international waters is less sure. The Sargasso Sea, where most of the sargassum is found, is vulnerable to exploitation at unsustainable levels because it resides within the political realm designated the high seas. This valuable, irreplaceable habitat needs protection by all concerned parties to ensure that the "weed" that Columbus observed more than five hundred years ago continues on its own long journey throughout the Gulf Stream and the Sargasso Sea.

From the minuscule plankton, which form the first link in the food chain of the giant bluefin tuna, the subject of the next chapter, to large aggregations of fish-holding sargassum, which big game anglers (Chapter 6) seek out, these drifters and floaters play an invaluable role in the vitality of the Gulf Stream.

5 BLUEFIN TUNA

THE GREAT MIGRATION

A FOUR-HUNDRED-POUND bluefin tuna in search of an easy meal cruises the nearshore waters off Cape Hatteras. The bluefin has ridden a warm-core ring that spun off from the Gulf Stream. Though some of the tuna's cohabitants in the ring may fall out into the cooler surrounding water, it resides within the comfort and feeding area of the ring, but its stay will be brief.

The bluefin tuna (*Thunnus thynnus*) is a nomad that wanders the length and breadth of the Atlantic Ocean. Long before the Spanish conquistadors and the New World settlers sailed upon ocean currents to reach their destination, bluefin used the Gulf Stream as a conduit for their own migrations. The Greeks were probably the first to apply the name "tuna" — meaning "to rush" — to these highly mobile fish. For more than four thousand years, philosophers and artists have sought inspiration from one of the most magnificent creatures navigating the warm waters of the Gulf Stream.

Aristotle recognized the migratory nature of this species and conjectured that bluefin tuna made annual spawning runs from the Atlantic Ocean, passing through the Strait of Gibraltar, to the Mediterranean Sea. For thousands of years, coastal villages all along the Mediterranean trapped bluefin tuna as they migrated through the local waters at the same time each year. The most famous of these ancient rituals is the *mattanza*, occurring during May and June in the waters off Sicily. The local fishermen, *tonnarotti*, construct and set a complex series of huge traps, *tonnaras*, which are composed of numerous passageways, chambers, and gates to ensnare the tuna. And the *tonnaras* are deadly efficient, funneling and herding tuna together for their eventual slaughter. Though now practiced mainly by Sicilian fisherman, the *mattanza* has its roots in Spanish and Arab culture. The term *mattanza* comes from the old Spanish word *matar*, "to kill." Just as the bull is brought down by the matador, the tuna trapped in the last room of the *tonnaras*, called the chamber of death, are killed one by one.

The Phoenicians and Carthaginians also prized the bluefin as a food source, and they appropriately paid homage to the fish by stamping their coins with an image of it. While this theme of bluefin tuna as a revered delicacy would be repeated in the twentieth century, the Greek philosopher Plutarch (46–120 A.D.) took a position on the tuna that deviated markedly from the gastronomical perspective of his fellow Mediterranean inhabitants. He wrote extensively on many moral and social topics of his time, including an in-depth treatise on whether terrestrial or aquatic organisms were smarter. In this tome, he described the schooling behavior of tuna as a conscious act that expresses their affection for one another, assigning almost human qualities to them. Pliny the Elder, in the first century A.D., championed the use of various parts of tuna as homeopathic remedies for a host of human ailments. In more recent history, Arabs, Spaniards, and Italians have integrated bluefin tuna not only into their cuisine but also symbolically into their poetry, music, and culture. Because of all this attention over the centuries, bluefin tuna have attained almost a mythic position among the world's pelagic fish.

Although bluefin tuna are widely distributed throughout the Atlantic basin, they have only two confirmed spawning locations: the Gulf of Mexico and the Mediterranean Sea. During the spring, female bluefin shed their eggs into the warm, open waters of the Gulf of Mexico to be fertilized by the attentive males. And these females are fecund; a six-hundred-pound tuna can produce more than 30 million eggs during one spawning episode. Like self-sustaining biospheres, these eggs contain an oil globule for nutrition and fresh water for buoyancy. From this encapsulated environment, tiny larvae emerge approximately two days after fertilization. Without the protection of any nest, the larvae are swept and scattered by the Loop Current across the sea surface. Survival in this exposed environment necessitates rapid growth. At a week old, the tuna are about a quarter-inch long, with disproportionately large eyes, mouth, and teeth, which serve them well as they begin their search for food. But in their quest to survive, the

Bluefin tuna (courtesy of NOAA)

young tuna must run the gauntlet of avian and aquatic predators. Losses are high. In the first two weeks after hatching, 50 percent of the larvae have succumbed to the rigors of their new pelagic home, and the losses continue to mount with time, approaching 99.9 percent after one year.

Moving with the Gulf Stream, as if drawn by some invisible force, the surviving young fish and their adult counterparts begin their journey northward through the Straits of Florida. As the weeks pass, they converge on the feeding and nursery grounds from Cape Hatteras to Cape Cod, arriving in late June to early July. Young bluefin make considerable weight gains during the summer, almost doubling in size from June to September. For the adult bluefin, the act of spawning is very strenuous. Females dispel the contents of their ovaries, weighing up to fifteen pounds, and males expel clouds of sperm into the water. The tuna must compensate for the energy expended in procreation by increasing their consumption of food. Feeding throughout the summer, these tuna prey upon a buffet of herring, mackerel, hake, bluefish, anchovies, and other creatures that roam the edge of the Gulf Stream. A mature tuna, weighing 450 pounds in June, may increase in weight to more than 600 pounds by the end of the main feeding period in October. Biologists believe this prodigious display of eating leads a fish to grow faster, mature sooner, and become larger. The net result, which biologists call fitness, is that these large fish produce more offspring per lifetime. Fitness is the foundation of natural selection: favorable traits that are passed on from generation to generation. As we will explore in depth later in this chapter, these traits, such as sustained swimming, ability to navigate in a featureless ocean, and thermoregulation, have made the tuna highly adaptable to the rigors of long-range migration in the Gulf Stream.

Bluefin tuna school and migrate from the spawning grounds according to size. Young bluefin ("schoolies") may range in length from approximately twenty to forty inches and attain weights up to 130 pounds. Medium bluefin tuna grow to lengths of sixty to eighty inches, approach 300 pounds in weight, and reach sexual maturity by the age of five to six years old. Giant bluefin tuna, which have a minimum length of eighty-one inches and a weight of more than 310 pounds, are the heavyweights of this group. And true to their name, these giants can reach weights of more than 1,400 pounds. It should be no surprise that the tuna must feed often and in great quantity to attain these enormous weights. Just how much food is necessary to support the bulk of these giant bluefin? As top-level predators, bluefin tuna are inextricably linked to even the lowest-level organism on the food chain — all flesh is grass. Biological principles dictate that a given food level passes only a small percentage of its energy on to the next level because most of it is lost or utilized at each step. The transfer of energy in the tuna's food chain is very inefficient, averaging only 10 percent. This low efficiency does not bode well for the giant bluefin, which must replace the weight lost during spawning and increase its fat content to weather the lean periods. Calculation of the weight gain based on 10 percent efficiency is relatively straightforward. For a tuna to gain ten pounds, it has to eat the equivalent of one hundred pounds of mackerel or other fish. But this interdependence does not end there, because the food chain of the bluefin may be as many as five levels long. Descending the food chain, it is estimated that it would take a thousand pounds of silversides to feed the mackerel, ten thousand pounds of small shrimplike organisms to feed the silversides, and one hundred thousand pounds of tiny plant matter for the shrimp. Though this calculation has a number of simplified, underlying assumptions, it is illustrative of the amount of food that the sea must produce to meet the nutritional needs of these large fish.

With large appetites to satisfy, these giants wander far and wide in search of nourishment. From the Gulf of Mexico, giant bluefin may travel more than two thousand miles along the Atlantic coast in about three weeks. Upon exiting the Straits of Florida, bluefin move along the western edge of the Gulf Stream, and as if on cruise control, they swim almost continuously in search of their next meal. Giant bluefin are opportunistic feeders, feeding whenever and wherever food is available. But throughout the vast expanse of the Atlantic, the distribution of food is spotty, lacking the dense concentrations of organisms to support the metabolic needs of these giants. In the seemingly barren landscape of the open ocean, the thermal fronts along the western edge of the Gulf Stream are plant-rich due to the infusion of fertile Slope Water. The plethora of prey organisms that feed upon the plant material soon find themselves in turn being targeted by the hungry tuna. By patrolling along these boundaries, tuna not only satisfy their need to feed but remain in a comfortable temperature zone.

This motivation to find food is so strong that tuna coordinate their movement and presence throughout the North Atlantic in response to the localized concentration of food. Schools of bluefin may depart the Gulf Stream, where the current veers to the east off Cape Hatteras, to forage in the coastal waters north of the Carolina coast. Others that have left the Gulf of Mexico will continue to ride the Gulf Stream, following the general northward migration of prey during the summer months. These tuna will congregate in the prey-rich waters off New England, where the Gulf Stream brushes past the Georges Bank. The tuna will have timed their arrival to coincide with the presence of vast schools of spawning herring, the keystone prey of the Georges Bank. This shallow submarine plateau is biologically quite productive, supporting an abundant array of marine species. This cornucopia of organisms owes its existence to the confluence of the cold, nutrient-rich Labrador Current, sweeping over most of the Georges Bank, and the warmer Gulf Stream. The mingling of the two currents, in conjunction with sunlight penetrating the shallow water, fosters an ideal setting for the explosive bloom of tiny phytoplankton. These minuscule plants grow three times faster on the Georges Bank than on any other comparable submarine feature along the Atlantic coast. Zooplankton graze upon this vast ocean pasture of phytoplankton. In turn, the zooplankton attract higher-level organisms, including the herring. And it isn't long before the voracious bluefin make their presence known to the unsuspecting herring. The water soon boils with activity as tuna feast upon the herring, and gannets, gulls, and other seabirds swoop down to clean up the remains.

The bluefins' quest for optimal feeding grounds may lead to some fish traveling ever farther from the Georges Bank. Migrations via the northern limb of the North Atlantic gyre often occur into the waters of Newfoundland, Iceland, and Norway. During the long summer days, these waters teem with life, providing a boundless supply of food for the hungry tuna. The aforementioned Iceland-Faroes Front is a particularly fecund area of the North Atlantic. The convergence of the Gulf Stream with the nutrient-laden Arctic waters sets the stage for a profusion of planktonic organisms, supporting a multilevel food chain that terminates with the tuna as the apex predator. Similar to the African nomads who for centuries crossed the vast sea of dunes of the Sahara Desert by a handful of routes, linked by a string of reliable wells, bluefin tuna depend upon the "oases" of the Gulf Stream to provide nourishment for their sustained journey.

Though already having traveled thousands of miles from the spawning grounds of the Gulf of Mexico to the northern reaches of the Atlantic, the journey of the bluefin continues southward via the Canary Current. As this current sweeps past the Strait of Gibraltar, some of the bluefin (estimated at 3 percent of the population) will enter the Mediterranean Sea to mingle with their eastern cousins and to be "recruited" into the Mediterranean aggregate. But the others move on. Upon reaching the North Equatorial Current, the bluefin travel across the tropical Atlantic to winter over in the warm waters of Brazil. As spring approaches, they head back to the Gulf of Mexico via the Caribbean Current. Those fish that completed the loop around the North Atlantic will have traveled more than ten thousand miles, essentially a nonstop journey on the oceanic gyre.

Many aquatic species, both freshwater and marine, are rheotactic—their movements are in response to the flow of a current. Trout will orient themselves to face upstream into the river flow, and bluefin, whether schoolies or giants, prefer to migrate by means of favorable ocean-rivers, like the Gulf Stream. How these fish navigate across thousands of miles of open ocean is still a mystery to researchers, but the current thinking points to a "pineal eye" on top of their head. This tiny part of the bluefin's anatomy is a thin piece of transparent skin like a window into their brain. In conjunction with the photoreceptors contained within the pineal organ, this "third eye" assists in the processing of light cues, such as the position of the sun. When navigation by the sun is impossible, bluefin may rely on a number of other environmental factors: the relative position of thermal fronts, the electromagnetic fields generated by a flowing current, and scent trails. In particular, bluefin tuna appear to have a strong affinity to horizontal temperature gradients. In order to confirm this idea, Barbara Block of Stanford University tagged twenty-eight tuna off the North Carolina coast. After leaving Hatteras, twenty-five of them traveled north along the Gulf Stream's western edge. Block was not completely surprised by this outcome because the Gulf Stream exhibits the greatest horizontal temperature change of the four currents in the North Atlantic gyre. In equatorial waters, where sea-surface temperature gradients are weak or nonexistent, bluefin tuna must switch sensing modes to navigate. As observed by Hatteras fishermen, bluefin tuna can locate chunks of oily bait, such as menhaden, from long distances, signifying a keen sense of smell. Scientists have postulated the possibility that bluefin tuna use the chemical signature of an ocean current to navigate, similar to a salmon returning to its natal stream by detecting the stream's particular brew of chemicals.

As opposed to Plutarch's hypothesis about bluefin tuna demonstrating a fondness for their neighbor and forming socially oriented schools, giant bluefin tuna appear to exhibit a rigidly defined school structure governed by their migratory and predatory rituals. Aerial surveys show one of the most common schooling structures is a parabolic formation — an arc usually limited to fifteen individuals and approximately two hundred feet wide. While many fish species congregate in schools that exhibit synchronous movements, the parabolic formation may be the ultimate utilitarian geometric shape in the myriad of schooling possibilities. Brian Partridge and his colleagues at the University of Miami discovered bluefin parabolas are functional on two levels: first, the position of the fish in the school is such that individuals benefit, conserving energy, from the hydrodynamical interaction with their neighbors. Similar to porpoises riding the bow wave of a ship to gain some forward momentum, a tuna may "push off" the pressure wave generated by its neighbor, gaining increased thrust through the water. Second, with the need to feed paramount among these migrating tuna, the curved formation can wrap around a ball of baitfish to prevent escape by the prey. This type of coordinated hunting and feeding, though observed in certain marine mammals (whales and porpoises) and in apex terrestrial carnivores (wolves and lions), is probably unique to the bluefin tuna in the entire piscatorial kingdom.

Though cooperative swimming and predation is a distinct advantage to a school of tuna, individual tuna have specialized anatomical, physiological, and biochemical traits that act and interact to govern their movements and distribution in the Gulf Stream. With 160 million years of unparalleled development in the tuna world on the bluefin's side, this species has readily adapted to its pelagic environment.

The rigors of long migrations necessitate that the bluefin tuna be proficient at propelling its body, which is immersed in a relatively high-density fluid. The bluefin must overcome the frictional forces of the water in order to achieve maximum efficiency in locomotion. One of the main frictional effects in the marine environment is the pressure drag, which is a function of the volume of water that must be pushed aside by a moving body. How does the tuna minimize the speed-sapping effect of the water? For that answer, we must look at the morphology of the tuna.

For many scientists and sportfishermen, the tuna represents the pinnacle of marine engineering by Mother Nature. Tuna have a highly streamlined body — fusiform shape — tapering at both ends from a thicker midsection. The caudal peduncle, which is anterior to the fish's tail, exhibits the greatest degree of tapering. The head is long and pointed, and even the eyeball is contoured to the body's surface. Endowed with this hydrodynamically efficient body, which results in a smooth, uniform flow of water over its surface and minimizes the formation of a turbulent wake, the tuna swims almost effortlessly through the fluid environment. If a tuna needs a burst of speed to chase down a food item, it can also retract its dorsal, pelvic, and pectoral fins into grooves and slots along its body. Rudderlike finlets, lying along the top and bottom sides of the caudal peduncle, further enhance the tuna's locomotion. As the fish moves forward, these finlets generate little vortices, similar to the dimples of a golf ball in flight, which contribute to the reduction of turbulence.

Though a streamlined tuna is blessed with a hydrodynamic, drag-reducing shape, the bluefin must also possess an efficient "motor." Propulsion in a tuna is a complex interplay between body movements and the caudal fin (tail). The basic movement of most fish is an undulating, wavelike motion from the head to the tail. The rate at which these waves travel along the body determines the efficiency of the forward movement. Fish with flexible bodies, called "carangiform" swimmers, generate considerable drag. In contrast, the rigid body of the tuna, a "thunniform" swimmer, produces a minimum of energy-draining waves that would impede motion. The sickle-shaped caudal fin, slicing through the water at an eye-blurring twenty beats per second, is the only noticeable movement of the bluefin. The alternating sweep of the caudal fin exerts a significant back pressure that provides the forward thrust of the fish.

The powerhouse to generate this force resides in the tuna's highly advanced red muscles. These vascular tissues contain high concentrations of myoglobin, a red pigment with a strong affinity for oxygen. The myoglobin delivers the oxygen to cell structures called mitochondria, which biologically manufacture a chemical (adenosine triphosphate) that acts as high-octane fuel for the cells of the muscle tissue. A bluefin's red muscle tissue may contain as much as ten times the amount of myoglobin found in other species. The marathon swimmers of the Gulf Stream need these red muscles, which operate nonstop, for prolonged swimming and other forms of endurance. While most fish have some red muscle tissue running in a narrow strip along the sides and just underneath the skin, bluefin have a large mass of this tissue occupying a broad internal band. The red muscle, inextricably linked to the caudal fin, provides the expansions and contractions that propel the tuna into its cruising mode.

A specific style of swimming, such as prolonged cruising or rapid acceleration, requires a particular caudal fin, which comes in a wide variety of shapes, sizes, and stiffness. A useful function for assessing the swimming style of a fish is the aspect ratio, a measure of the height of the caudal fin in relation to its surface area or broadness. There is a positive correlation between efficient, top-end cruising and high-aspect ratios. Why is this so? A caudal fin that is broad can develop a significant push against the water for jackrabbit acceleration or maneuverability. Demersal fish (a flounder is a representative example) that live near or on the seafloor employ both camouflage and stealth to ambush their unsuspecting prey. When the opportunity presents itself, they spring from the bottom, powered by their broad tail, to engulf the prey. But these sedentary fish could not adapt to the open waters of the Gulf Stream because the downside of a broad tail is increased drag, resulting in loss of sustained motion. In contrast, riders of the Gulf Stream have caudal fins with minimum surface area but maximum spread between the dorsal and ventral lobes of the fin. These fins are indispensable for long-distance cruising fish. Like a car spinning its wheels, these fish are slow to get going, but once they get moving, they can maintain relatively high speeds due to decreased frictional drag. Bluefin tuna that have been released back into the water by Hatteras fishermen will beat their tails furiously, but there is initially little accompanying movement. It's as if these fish were futilely swishing their tails to obtain purchase on the water. Since the bluefin is the apex cruiser of the Gulf Stream, it has a lunate (crescent-shaped) caudal fin, maximizing the aspect ratio. The fin is very rigid and relatively useless for maneuverability but effective in maintaining top-end speeds. The lunate fin, which is attached to a flexible caudal peduncle and powered by the red muscle tissues, makes the bluefin a formidable feeding machine, overtaking prey with relative ease. But specialization in one direction requires tradeoffs in other areas. A bluefin tuna can cover large stretches of the watery realm, but because of its limited turning ability, it fails in capturing a significant number of baitfish from the pods it encounters. In order to satisfy its nutritional needs, the hunt for prey and the migration to food-rich waters of the Gulf Stream must continue almost nonstop. The bluefin is genetically programmed to be a marathon swimmer, not a sprinter.

To biologists, the process of metabolism is of the highest importance to all life on this planet. It is not a single process but a complex of chemical and physical processes of the body that provide the energy for growth and maintenance. Bluefin tuna, at all sustainable swimming speeds, have metabolic rates markedly above those of other similarly active fish. High metabolic rates require significant amounts of oxygen as the "fuel" to convert food to energy. Though water is a molecule of two atoms of hydrogen and one atom of oxygen, no marine animal has the ability to break down the molecule to obtain its oxygen directly. It must rely for its supply of this vital gas on either the production of oxygen by plant photosynthesis or by exchange with the overlying air. The Gulf Stream "breathes in" oxygen as it diffuses gradually from the atmosphere across the air-sea interface to the water. At times, the Gulf Stream "breathes deeply" as wind mixing and turbulent water allow more oxygen to be stirred into the water. But similar to a warm soda going flat, the current "exhales" oxygen back to the atmosphere. The net result is that seawater contains very little oxygen, only about 3 percent of the oxygen that the atmosphere contains. Even more distressing for the bluefin, warm water has less dissolved oxygen than cold water, so the Gulf Stream may contain 40 percent less oxygen than the Labrador Current. But the bluefin is a highly adaptable animal; it can efficiently extract even this meager amount of dissolved oxygen. The gill surface area, over which the dissolved oxygen passes, is very large in tuna — the largest among all fishes. The size of the total respiratory area approaches that found in the lungs of mammals of comparable weight. To supply its gills with oxygenated water, the tuna must remain in a state of perpetual motion. Having lost the ability to actively pump water over their gills, bluefin tuna literally swim with their mouths open, allowing for the flow of water over their gills. Known as ram ventilation, this mode of respiration is mandatory for the survival of the tuna — if they stop swimming, they suffocate. Suffice it to say, bluefin are hardwired to cruise the watery environment.

Temperature, in part, controls an organism's metabolic rate, so an organism that is able to regulate its internal temperature through a wide range of ambient temperatures is conferred a significant survival advantage. During its great migration around the North Atlantic gyre, the bluefin tuna swims immersed in ocean currents with temperatures ranging from a chilly 50° of the Labrador Current to a warm 80° of the Gulf Stream. And these temperature extremes can occur over a relatively short distance, as when the bluefin patrols along a thermal front for food. In general, metabolic processes accelerate with increasing temperature, so an organism that is able to internally generate heat will be more active, be better able to chase prey, and have higher rates of digestion. Most fish are "cold-blooded," or ectothermic, having an internal body temperature similar to that of the environment. In contrast, bluefin tuna are able to maintain elevated temperatures even when swimming in the cold northern waters during their migration. These "warm-blooded," or endothermic, organisms can pursue fast prey, such as mackerel and squid, because the flow of blood into muscles warms these tissues, resulting in increased efficiency. Specifically, muscles will contract three times faster when they have been warmed ten degrees.

But how does the tuna generate its heat and maintain its relatively high internal temperature? What type of heating system does the tuna possess? When the tuna metabolizes food, it produces heat as a by-product, which warms the blood a bit before it is shunted back to the gills. But this heat is rapidly lost to the relatively cool surrounding water that flows over the gills. The blood that returns to the muscles is as cold as the ambient water. The tuna could generate more heat by increasing its metabolism, but the downside is that increased metabolism requires more oxygen, a greater flow of blood across the gills, and increased heat loss at the gills due to the high thermal conductivity of water. The fact is an animal with gills cannot raise its body temperature solely through metabolism.

The tuna elevates its body temperature through a complex of fine, parallel arteries and veins: the *rete mirabile* (Latin for "wonderful net"). This network functions as an efficient countercurrent heat exchanger. As warm, oxygen-poor blood passes through the veins in the rete on its way to the gills, its heat is transferred to the adjacent, cooler arterial blood, coming in the opposite direction from the gills. As this arterial blood passes through the entire network, its temperature increases and supplies the necessary heat to warm the tuna's muscle. In this unique display of thermoregulation, little heat is lost to the environment, and most of it, 98 percent, is stored within the animal's body. Thus, the tuna has unique physiological advantages: efficient cruising speeds, expanded geographical range, and increased foraging opportunities — over its cold-blooded counterparts.

The high metabolic rates that are found in tuna require a high rate of blood flow to their tissues. To understand how this is done, one has only to look at the cardiac physiology of the tuna. Tuna hearts are larger, beat faster, and pump more blood than the hearts of other fish. Marvelous beasts of natural engineering, bluefin tuna have all the physical and biological "tools" to successfully roam the seas, as they have been doing for millions of years. They have no peers in the piscine world when it comes to riding the North Atlantic gyre.

Aristotle was cognizant of bluefin migration only in the Mediterranean Sea; he was most assuredly unaware of the transatlantic wanderings of these muscular torpedoes of the sea. Ignorance of the migratory habits of bluefin tuna would persist for centuries after Aristotle's time. This lack of knowledge frustrated scientists, who had many unanswered questions about the bluefins' open-ocean movements, stock structure and stock mixing, and reproductive biology. The mid-twentieth century marked the beginning of the first systematic attempts by fishery scientists to track these animals. While initial interest in the wanderings of these creatures arose out of basic scientific curiosity, conservation efforts to save the species from overexploitation by commercial fishermen mainly drive today's tracking efforts.

In May 1952, biologists Frank Mather and Howard Schuck attached the first "tags" (stamped and numbered hooks) to bluefin off the island of Bimini. Though these conventional tags would become the preferred method for determining the movement of bluefin, the researchers and their assistants had to recapture the fish in order to remove the tag — a significant drawback in the employment of this method of tracking. While Mather and Schuck were ecstatic that one of their tagged giants had been landed in Nova Scotia, they soon discovered that the return rate on these tags was less than 10 percent. In addition, since the exact path traveled by the fish between its tagging and recapture points could not be assessed with any confidence, Mather and Schuck had to be content to receive only limited spatial information on the tuna's movement. But over the decades, the indefatigable Mather had collected enough data on bluefin tuna to challenge some long-standing opinions on the biology of these fish. In particular, he rebutted the prevalent belief that the only bluefin tuna spawning ground in the western Atlantic was in the Gulf of Mexico. While he acknowledged some tuna swim through the Caribbean Sea and spawn in the gulf, he argued that the main body follows the Antilles Current and spawns east of the Bahama Islands, a long way from the Gulf of Mexico. To the skeptical scientific community, Mather and Schuck's tags could never resolve this issue, and there was much more that they wanted to know. A new technology was needed.

Acoustical tags track the movements of fish by means of receiving and processing the sound pulses emitted from a small transmitter attached to the fish. Although acoustic tags have been used to monitor fish movement for more than thirty years, Frank Carey of Woods Hole Oceanographic Institution was a pioneer in the use of these tags on large pelagics, such as blue sharks and swordfish. The tags that Carey employed yielded information only on the gross spatial patterns of the fish and had a limited signal range, requiring the tracker to remain in relatively close contact with the fish. More recently, acoustic tagging techniques have tracked the wanderings of bluefin tuna on a finer scale, particularly from a three-dimensional perspective. Bluefin tuna in the Gulf Stream may sound repeatedly, searching for their optimum foraging depth. Tuna often go diving after squid at dawn and dusk and remain at depths of twenty-four hundred feet. A bluefin tuna can be basking in the warm surface waters of the Gulf Stream at one moment, but the next instant, with a swish of its tail, be swimming in the cold abyss.

In the middle 1990s, two different paths would converge, opening new frontiers in the tracking of bluefin tuna. First, the rapid pace in the miniaturization of computer chips would foster the development of microprocessors that could store large amounts of data (two megabytes) in a tiny package. Second, recreational fishing for bluefin tuna exploded in popularity when commercial fishermen discovered in the winter of 1994 a massive concentration of these fish inhabiting shipwrecks about twenty miles from Hatteras Inlet. Some of the bluefin that had summered over in the northern feeding grounds moved south to the warmer waters adjacent to the Gulf Stream, possibly reflecting a new migration pattern. It wasn't long before the Hatteras charter fleet would go after these bluefin. The winter pall that normally settled on the tiny village of Hatteras lifted, and things would never be the same for the town and for me.

I remember being cold, not just a nagging discomfort, but downright miserable, as the cold seeped deep into my core. But I had plunked down more than a thousand dollars for the bluefin tuna charter, so I was determined to block out the pain. (Yes, more than a grand to fish. I long ago stopped rationalizing my obsession with fishing to nonanglers, having accepted the philosophy of the author John Hersey who, when attempting to explain his passion for bluefish to a stranger, simply says, "Fishing is complicated.")

Previous fishing reports spoke of a school of giant bluefin tuna that had taken up temporary residence over a wreck south of Hatteras Inlet. With the coordinates in hand, we steamed in that direction, making good time, thanks to a glass-slick sea. Upon approaching our destination, I observed wispy plumes of "smoke" gently rising from the sea surface. It was as if the sea itself were smoldering. An omen? Some Icelandic sagas — narratives of Norse history and legend dating back to the twelfth century — speak of hideous sea serpents lurking below the surface, belching fire and smoke. No sea monster for me, I had witnessed a markedly visible sign of air-sea interaction — steam fog, or sea smoke, a common phenomenon at these latitudes during the winter. As cold air (when we left the boat's slip, the temperature was 17°F) flows over the relatively warm water, heat and water vapor ascend from the sea surface. Mixing with the cooler air above, the water vapor condenses into small droplets, which makes the plumes visible, similar to your breath on a cold day. To any casual observer, they give the appearance of distinct rising columns of smoke. I took it as a good sign.

Once we were positioned over the wreck, the mate, Sig, began cutting menhaden and other baits into chunks, chum to stimulate the bluefin's hunger juices. In the blink of an eye, his knife slipped, slicing his index finger, blood spurting from the wound. Not a good sign. I girded myself for the stream of profanity that I expected to flow from Sig. Nothing. His hand was so numb from the cold that I believe his senses had gone into hibernation. Patched up, he ladled the bloody concoction overboard, a nice slick coating the surface.

From the depths, a large bluefin rocketed to the surface, leaping clear out of the water, with its massive maw agape. Three others quickly followed, all intent on getting their share of the floating morsels. The bite was on, feeding time in this watery zoo. For a minute, I just stood back, staring in disbelief at this display of unbridled kinetic energy. My trance was broken when Sig began to strap me into a fishing harness. Called the "brute tamer," it resembled some contraption to control Hannibal Lecter in the film *Silence of the Lambs*. The harness, when used with the proper technique, allows the angler to fight a bluefin standing up and apply maximum pressure on the fish. I opted out — no way was a determined bluefin going to pull me over the boat's stern — and settled into the fighting chair. Sig impaled a chunk of menhaden on the hook and tossed it over, and I was soon hooked up to a runaway bluefin headed straight to the bottom. Twenty-five agonizing minutes elapsed before I could turn the fish and start to retrieve line. Like a crippled submarine, the bluefin slowly rose to the surface; its moon-shaped eye staring back at me. Sig was quickly on the fish, using a sharp needle to insert a "spaghetti tag" (named for its resemblance to the noodle) through the skin, just under the dorsal fin of the fish's back. With the circle hook removed from the corner of the bluefin's jaw, its freedom was now assured, and none too soon for my aching arms.

With the cooperation of sportfishermen willing to aid in the capture of these fish through hook-and-line fishing and top-shelf technology at hand, Barbara Block of Stanford initiated the Tag-A-Giant (TAG) program. As tuna were brought on board the boat, Block and her colleagues would surgically implant an archival tag, microprocessor-based, into their guts and quickly release them. Not only did the tuna survive their ordeal but the tags recorded and stored the geo-position (based on ambient light levels), thermal physiology, and diving behavior of these fish every two minutes. Although the researchers hoped to receive a bonanza of data from the tagged fish, a major limitation to archival tagging, like conventional tagging, is the need to recapture the fish to access the data. This meant that the participants in this joint effort had to catch, tag, and release a large number of fish to be assured of a significant number of returns, which might take years to retrieve. In addition, the multinational nature of the bluefin, a by-product of its extensive migration, complicates the coordination and recovery of archival tags. At present, the researchers have been disappointed with the small number of tags that have been recovered for analysis. But of the eight archival tags recovered, one was attached to a fish that had traveled well into the Mediterranean, further demonstrating that bluefin tuna do migrate into the eastern Atlantic via the Gulf Stream.

In 1992 and 1993, Frank Carey ushered in a new era of tracking large pelagics by successfully attaching satellite tags to the dorsal fins of blue sharks. When the sharks are cruising near the surface, the opportunity arises to uplink the data from the tags to orbiting satellites. While Carey was the first to demonstrate the use of satellite tags on game fish (marine mammals and birds had previously been outfitted with satellite tags), he also realized they were limited to those fish that frequent the surface. Recognizing the potential of remote sensing, Block and her associates took the technology that Carey introduced to the next level. They designed a pop-off tag that is attached externally to a fish, releases ("pops off") at a programmed time, and surfaces to begin the transmission of data to the satellite.

In the winter of 1997, Block's group returned to Hatteras, eager to test the pop-off tags in the TAG program. With the able assistance of Bob Eakes, who has hundreds of certified bluefin catches to his credit, they tagged thirty-seven fish in the 250–300-pound range. Block outfitted nine of these fish with tags that she had programmed to release relatively soon after the fish were set free. During this period, the western edge of the Gulf Stream was close to Hatteras and formed a distinct edge with the Labrador Current to the north. The temperature change across this thermal front was more than twenty-five degrees in one nautical mile. All nine tags popped off in the Gulf Stream or along the frontal zone, indicating the bluefins' preference for the warmer water mass to the cooler Labrador Current or coastal water. The remaining twenty-eight fish received long-term tags, programmed to deploy after a sixty- to ninety-day time frame. One of the objectives was to examine the movement and dispersion of these fish in the western North Atlantic. All the tagged fish, except one, traveled northeast of their release point and along the path of the Gulf Stream. Block also noted that bluefin can move relatively rapidly into the eastern Atlantic, as one did. This tuna traveled 1,670 miles in ninety days, living up to its species' reputation as highly migratory. But this migratory nature would spell trouble for the tuna.

During the first half of the twentieth century, commercial fishermen exerted little pressure on the western Atlantic bluefin tuna population. With a dockside value of a few cents per pound for tuna, fishermen had little economic incentive to pursue these fish. Only New England fishermen ventured out to sea to catch tuna by hand gear or nets and sold their limited catch to cat food and dog food producers. But the latter half of the twentieth century would usher in marked changes in the bluefin tuna fishery: development of mechanized purse seine fishing in the late 1950s, implementation of high seas long-lining by the Japanese, and increased demand for bluefin tuna by Japanese sashimi markets in the 1970s. As Japanese fishing fleets ranged throughout the oceans in search of bluefin, they soon discovered that the Mediterranean and New England fish had the prized higher fat content than their Pacific counterparts. This fresh, fatty flesh could command top dollar for the seller. The equivalent of the California gold rush began in the 1970s as Japanese and, now, American fishermen actively scoured the sea for these fish. There was money to be made. With the development of fast airfreight, fresh giant Atlantic bluefin could reach Japan's lucrative sushi and sashimi markets overnight. In 1973, Japanese importers saw a jump in the price of tuna from five cents per pound to more than a dollar per pound. By the 1970s, the western Atlantic bluefin tuna population exhibited signs of stress from overfishing. To limit the impact of foreign fleets on domestic stocks, Congress passed the Magnuson Fishery Conservation and Management Act in 1976 (renamed the Magnuson-Stevens Fishery Conservation and Management Act in 1996). Ironically, this law did nothing to protect the beleaguered bluefin tuna. The lawmakers argued that the highly migratory nature of the bluefin precluded the implementation of effective management strategies and the establishment of jurisdictional zones.

Japanese buyers in 1986 were paying $12 per pound for tuna caught in American waters by Americans. (The implementation, three years earlier, of the EEZ, which excluded all foreign fishing vessels without special permits, essentially replaced foreign overfishing with domestic overfishing.) In 1991, a single giant bluefin sold for a record price of $68,503, about $100 per pound, to a Japanese buyer.

Recognizing the need for coordinated international cooperation on the migratory bluefin, more than twenty nations, including the United States, Japan, and most of the fishing nations in the North Atlantic, established the International Commission for the Conservation of Atlantic Tuna (ICCAT). In the early 1980s, ICCAT proposed managing the Atlantic bluefin tuna as separate east and west units, divided at the 45° west meridian. The ICCAT managers instituted this two-stock proposal based upon the assumption of two isolated spawning areas and relatively insignificant transatlantic mixing rates. Many in scientific circles viewed the 45° dividing line as nothing more than an artificial boundary. What highly migratory fish is going to abide by this manmade demarcation line? As scientists evaluated anecdotal evidence of bluefin migration and analyzed data from tagging studies, they came to the conclusion that mixing between the eastern and western stocks was a distinct possibility. To whoever would listen, the researchers argued that there was no biological evidence that indicates the large geographical separation between the tunas' two spawning areas represented reproductive isolation. Barbara Block and others are continuing to carry this message as they attempt to quantify the extent of bluefin migration between the western and eastern Atlantic. It may be that the salvation of this prodigious traveler of the North Atlantic gyre lies in its long imprinted instinct to wander.

6 FISHING THE BLUE WATERS

EARLY PRAISES FOR Gulf Stream fishing come from no less an authority than Ernest Hemingway, recognized by the angling world for his development of innovative deep-sea fishing techniques and record-setting billfish catches. In a 1936 article for *Esquire* magazine, Hemingway argued that fishing in the Gulf Stream ranked above big game hunting on the African plains: "In the first place, the Gulf Stream and the other great ocean currents are the last wild country that is left. Once you are out of sight of land and the other boats you are more alone than you can ever be hunting and the sea is the same as it has been since men ever went on it in boats." Hemingway, who went on to win the Nobel Prize for Literature, would draw from his numerous fishing adventures in the Gulf Stream to craft his acclaimed novel *The Old Man and the Sea* (1954). In this epic tale, the old Cuban, Santiago, ventures into the Gulf Stream to do battle with a marlin, which after a prolonged fight he finally catches — only to watch it be ravaged by sharks. Throughout this literary work, Hemingway consistently calls upon his intimate knowledge of the Gulf Stream to create the dramatic clash of human and animal against the backdrop of this powerful current.

Game fish, including dolphin (the fish, not the mammal), wahoo, yellowfin tuna, and billfish, frequent the Gulf Stream waters during the spring to fall months. Yellowfin tuna arrive first, taking center stage from mid-March to May. With the advent of rising water temperatures during May, dolphin and billfish begin to appear and remain throughout the summer. Though an isolated wahoo can be mixed in with the above species, their numbers peak in September and October.

Lured by the possibility of tangling with one of these predators, anglers from all over the eastern seaboard and beyond descend upon the coastal hamlets of the Outer Banks of North Carolina. Probably nowhere is Gulf Stream fishing more a part of the history, culture, and economy of a town than in the tiny village of Hatteras. Known as the "billfish capital of the world," Hatteras sits on a spit of land that juts out into the Atlantic, allowing for a short run to the Gulf Stream by its fleet of charter boats.

The sport of offshore fishing in the Outer Banks began in 1937, when Ernal Foster, a longtime resident of Hatteras, conceived the idea of building a boat that could be used for commercial fishing in the wintertime and sportfishing the rest of the year. Though fellow fishermen ridiculed his proposal (pay money to go fishing — not a chance — the country was struggling to rid itself of the cloak of the Depression), Ernal was determined to fulfill his dream. His original boat, the *Albatross*, was to become one of three boats built between 1937 and 1953 that comprise the Albatross Fleet. The durability of these boats and the commitment of those who pilot them stand as a testament to the fact that this offshore fleet has fished every season except during the years of World War II. Ernal Foster's *Albatross* cost only $805 and made only four trips, at a price of $25 per trip, to the Gulf Stream during its first year, but blue-water fishing in the twenty-first century has changed considerably since Foster's time. While Ernal gathered wood, mostly juniper, and cured it himself to fashion the *Albatross*, the vessels of today sport fiberglass hulls and price tags approaching seven figures. To the Hatteras old-timers and even some not so old, these "plastic boats" can never replace the handcrafted wood vessels of a time long gone.

The Gulf Stream lured other authors before Hemingway's time, as well. Zane Grey, known mainly for his extensive chronicles of the American Old West, also wrote about fishing in the Gulf Stream. *Tales of Fishes* (1919) and *Fishing in the Gulf Stream* (1922) describe using light tackle to catch sailfish off the Florida Keys.

Following in the wake of Foster, Hemingway, and Grey, dozens of diesel-powered Carolina sportfishers, reaching lengths of more than fifty feet, leave almost daily from Hatteras Harbor, Teach's Lair, and Hatteras Landing marinas. Quite often these boats have on board a makeup charter — a maximum of six anglers, willing to share the cost of a twelve-hundred-dollar charter. Skill levels can range from novice to master angler, but all are willing to spend a day on the Gulf Stream pursuing their angling dreams.

Before heading east from Pamlico Sound to the fishing grounds of the Gulf Stream, the fleet must pass through Hatteras Inlet. Many inlets along the eastern seaboard can be treacherous to navigation, and Hatteras Inlet is no exception. The presence of shifting shoals, strong tidal currents, and breaking waves can test the mettle of even the most experienced captain. Caution is the operative word among Hatteras captains when attempting to run the inlet. Prudent captains will throttle their vessel down upon reaching the inlet, waiting for a lull in wave activity before running this watery gauntlet.

Upon exiting Hatteras Inlet, a trip to the Gulf Stream may last just over an hour to about two hours, depending upon the proximity of the current to the coast. Regardless of the duration, the boat's mate puts this time to good use, preparing bait and tackle for the upcoming task and instructing anglers on the proper angling protocol. Only the anglers are anxious, bursting with anticipation to tangle with the fish of a lifetime.

As the Gulf Stream draws near, one of the first signs of life of this unique ecosystem is the presence of flying fish, skimming above the water surface. Even novice anglers easily recognize this fish by its winglike pectoral fins and its lopsided tail with the lower lobe larger than the upper lobe. Chased by predators, such as tuna, a flying fish swims rapidly to the surface, propelling itself by means of its powerful tail into the air. The fish's unique combination of outstretched pectoral fins and asymmetrical tail permits it to gain height, extend gliding time, and remain airborne for more than three hundred feet.

The first published account (1791) of this species, according to Duke University historian Peter Wood, was by John Gabriel Stedman, who, while serving on a military mission in the tropical Atlantic Ocean, was privileged to catch numerous flying fish in the ship's shroud. These involuntary subjects would start Stedman on a long career of observing and studying these fish. He was to conclude that while flying fish leap out of their watery realm to escape one set of pursuers, that puts them at the mercy of another — predatory seabirds. As Professor Wood insightfully points out, "Stedman came to understand that mariners viewed this aquatic species as the epitome of those obligated between the devil and the deep blue sea." The "devil" was sailor's slang for a seam around a wooden ship's hull that extended all the way down to the waterline. These seams required periodic caulking with pitch to avoid leaking, and often this repair had to be done while the ship was at sea. A sailor who was lowered over the side found himself in a precarious position, much like the fleeing flying fish.

More than a century after Stedman's work, Winslow Homer picked up the theme of flying fish in his seascape, mentioned earlier, *The Gulf Stream*. The painter depicts a school of flying fish rising from the sea in a futile attempt to elude the thrashing sharks that are menacing a black man's dismasted skiff. The flight of flying fish became a metaphor for the slaves who attempted to escape the bondage of slavery during the nineteenth century.

Even today, the flying fish enters into the world of offshore fishing. Captains frequently employ "bird rigs"—lures that have a flying fish profile. When towed behind the boat, these lures provide visual and auditory cues to cruising game fish. Some enterprising boat captains in Florida have taken the mimicking of flying fish to the next level. They rig a dead flying fish with its wings pinned open and troll the offering at relatively high speeds under a kite. The kite line is repeatedly pulled down, jerking the bait out of the water and letting it fall back again, to simulate the flight of a live flying fish. Though time-consuming, if done correctly, this method is deadly effective at catching tuna.

A decrease in engine noise is a sure signal that it is time to go fishing. As if programmed, the mate springs into action, deploying lines, bait, and lures overboard to form a trolling spread. Like a spider web, these lines radiate to the port, starboard, and stern of the boat, where they are attached to various outrigger poles and clips that will hopefully keep them from tangling into a Gordian knot. So begins the often tedious but anticipatory ritual of dragging an assortment of teasers, skirted bait (having a feathered dressing placed over it) as well as naked bait, and squid chains behind the boat, with the intent of enticing a tuna, wahoo, or other pelagic species in search of easy prey.

To the untrained eye, it may appear that these lures and bait are skipping over a seemingly endless expanse of homogeneous water. But the key to success is spatial heterogeneity, observable or measurable changes in particular physical or biological properties along the current. Competent charter boat captains probably possess innate perceptual-cognitive skills. One striking example is the ability of the captain to "read" the water — observe nature, think simultaneously about temporal and spatial relations, and arrive at a rational conclusion. This ability unites the visual, intellectual, and conceptual aspects of fishing in a heightened feel for the physical place. Ernal Foster definitely had the "right stuff," catching record-size fish in waters off Hatteras that were thought by many to be barren and accomplishing these feats without the aid of today's navigational aids, such as depth finders, GPS, and autopilots. The "place" is at the essence of locating fish. A basic tenet of ecology goes something like this: organisms go where there is food. While this statement seems trivial, its applicability to successful Gulf Stream fishing is undeniable. Wherever there is a concentration of prey, hungry predators are present, ready to play their role in the marine food chain. To a Gulf Stream game fish, thermal fronts mean food, plenty of it. According to Mitch Roffer of Roffer's Ocean Fishing Forecasting Service, "If this food chain exists within a game fish's preferred temperature, it's only a matter of time before they find it and stack up in the area." Thermal fronts, especially those with well-defined edges, can act as a barrier to the movement of game fish. Seeking their thermal comfort zone on the warm side, the fish will move along these edges instead of crossing them.

Even subtle signs — a warm breeze coming across the bow of the boat, changes in water color — reinforce the captain's intuition that the trolled baits are in "fishy" water. Trolling is probably the most productive and favorite method employed by the offshore angler because at its core is the art of mimicking free-swimming prey. Ballyhoo, a close relative of the flying fish, makes excellent bait for all targeted species in the Gulf Stream, and if handled and rigged properly, it can be trolled longer and faster than live bait. The key to eliciting a strike from a hungry fish is for the ballyhoo to appear natural as it skips across the waves. To achieve that end, the mate often literally squeezes the feces out of the ballyhoo, making it more pliable, before inserting the hook through its cavity. Though a trolling speed of six to eight knots is the norm, if the current is flowing against a brisk wind, resulting in a heavy sea, slower speeds are in order to keep the bait in the water. Flexibility and adaptation by the captain to what the Gulf Stream offers up on any given day may be the difference between boating fish or telling fish stories back at the dock. In his *Esquire* article, Hemingway further expands upon the fickle nature of the Gulf Stream: "In a season of fishing you will see it oily flat as the becalmed galleons saw it while they drifted to the westward; white-capped with a fresh breeze as they saw it running with the trades; and in high rolling blue hills the tops blowing off them like snow as they were punished by it."

Once the trolling spread has been set in place behind the boat, the captain and the mate keep a watchful eye on it. To the untrained observer, the sea appears as a boundless blue mat, interrupted by the foam of whitecaps and the bubbly trail of the ballyhoo. But they are looking for a sign: the flash of color streaking across the spread that signals a game fish has taken an interest in their offering. Their simultaneous cry of "big dolphin" elicits a knee-jerk response from all the anglers, who now furtively scan the water surface. But this "scratching of the surface" doesn't do; one has to peer deeper. And this can only be accomplished by significantly eliminating reflected sunlight from the water surface. Our eyes are comfortable with the ambient light until the intensity of this light reaches approximately thirty-five hundred lumens—the unit of measurement in physics to determine the degree of luminosity. When the brightness of the reflected light reaches about four thousand lumens, the eyes begin to experience difficulty coping with the intensity, and bright areas appear as silvery-white flashes, commonly called glare. Knowledgeable offshore anglers would not leave port without a pair of polarized sunglasses, which permit only the vertical component of light to pass through; glare, being the horizontal component, is blocked out, improving contrast and increasing the angler's comfort level.

The first dolphin that flashed unscathed through the spread may be quickly followed by another, who is intent on inhaling the bait trolled from the left outrigger. The "popping" of the release clip frees the line from the outrigger pole and signals the start of fishing's version of a Chinese fire drill. Orders are barked to clear all lines (reel in the other trolled baits to avoid tangling with the hooked fish), man the fighting chair, and, at all times, point the rod tip in the direction that the fish is moving. If all on board can choreograph these tasks smoothly, then their reward is an exciting and acrobatic fight by these beautifully colored fish. Against the backdrop of the blue waters of the Gulf Stream, the dolphin's coloration is quite dramatic, a kaleidoscope of colors: iridescent greens and blues along its back, changing laterally though a green-gold-yellow spectrum along its flanks. But the dolphin's beauty is short-lived; its ephemeral colors quickly fade upon boating to uniform dark silver — a shroud signaling its death.

Known formally by its scientific name *Coryphaena hippurus*, the dolphin (or dolphinfish) was first described and classified by the famous taxonomist Carolus Linnaeus in 1758. Preferring warm water (78° to 85°F is optimal), dolphin can be found consistently from the Caribbean basin to the Gulf Stream waters off the North Carolina coast and as far north as New York. Though little is known about their migratory habits in the Gulf Stream, fishery scientists have hypothesized that they respond to seasonal changes in water temperature in search of more productive food sources. During their seasonal migrations, they travel in schools known as pods. The size and makeup of the pods are a function of the size and sex of the individual dolphins. Larger males (bulls) and females (cows) generally travel alone or in pairs. The mature males are easily distinguished by their high, blunt foreheads and are usually larger than the females, attaining weights of more than eighty pounds. In contrast, small fish (schoolies) travel in pods ranging in size from a few fish to several dozen.

Juvenile dolphin have a particular affinity to sargassum weed and other floating objects. Attracted to the smorgasbord of tiny marine organisms that inhabit the sargassum, dozens of dolphin may be found holding just below even the smallest sargassum clumps. The lure of floating objects — buoys, logs, and even dead marine mammals — for the dolphin cannot be overstated. The scientific jury is still out with regard to the appeal of the fish attracting devices (FAD), but food, shelter, and protection all rank high on the list. During the Cuban refugee crisis of 1994, thousands of Cubans attempted to flee the economically crippled regime of Fidel Castro. These desperate people launched all types of makeshift rafts, hoping to make the ninety-mile crossing to Florida. Though the coast guard rescued the majority of the refugees, thousands of abandoned rafts, which in many cases were no more than pieces of plywood, remained adrift at sea and some floated all the way to the Outer Banks — a 780-mile drift in the Gulf Stream. During the following year, anglers from Florida to North Carolina

Dolphinfish (copyright iStockphoto.com/Richard Gunion)

reported unprecedented catches in dolphin that had taken up residence under these rafts. A potentially catastrophic human tragedy had spawned a fishing bonanza.

Once an angler is hooked up to a small dolphin, the mate instructs the angler to keep the schoolie in the water in order to keep the attention of other dolphin holding under the sargassum. Seemingly drawn by an innate curiosity about the plight of their fellow schoolie, the fish move closer to the boat's stern. Chumming with cut bait, such as squid, the mate attempts to trigger the feeding instinct within these fish. With insatiable appetites to satisfy, the dolphin are more than willing to accept these free offerings. But once the dolphin shift into a feeding frenzy mode, the bait now comes with a hook. With up to four anglers in close quarters at the stern of the boat, anglers drift chunks of squid back to the waiting fish. So willing are these fish to eat that they literally hook themselves. With up to four fish hooked at once — darting and jumping in all possible directions — lines cross, choice words fly, and tempers flare. But a seasoned mate often saves the day — cajoling, instructing, untangling, and, most important, boating fish for a happy group of anglers.

Dolphin catches in the waters of the Gulf Stream are subject to federal regulations, which are established by regional fishery management councils. The South Atlantic Council, which includes the states of North Carolina, South Carolina, Georgia, and Florida, has presently established a maximum catch limit of sixty dolphin per boat. While this number may seem excessive, dolphin mature rapidly and are prolific breeders, making them an excellent species to sustain high catch levels without fear of overexploitation.

Dolphin may very likely be the fastest growing game fish known to the scientific community, increasing in size at the phenomenal rate of one-half to more than one inch in length per week. Based upon these rates, a dolphin can attain a body length of four feet and grow to forty pounds during its first year. Spawning occurs in the warm, open waters of the Gulf Stream from spring through the fall, but observational evidence suggests year-round spawning in the tropics. Dolphin are batch breeders, reproducing multiple times per year. Sexual maturity may occur as early as three months, when the dolphin is as small as fourteen inches, but all fish reach sexual maturity by the time they attain a length of twenty-two inches. But the good do die young — the maximum life span of a dolphin is only four years.

To support their rapid growth, dolphin must feed often and in quantity. But they are not particularly selective in choosing their prey, feeding on small pelagic fish, juveniles of larger pelagic fish, and invertebrates.

After one particularly productive fishing trip to the Gulf Stream, I became acutely aware of just how eclectic the diet of dolphin can be. Fish that are brought back to dock are generally cleaned by the staff of the marina. I am always amazed at their skill in filleting a forty-pound dolphin with one slice of their knife, albeit a very sharp one. I asked for the carcass of a filleted dolphin. "Sure, no problem," mumbled a blood-covered cleaner. "Are you going to use it for crab bait?" "No. I wish to see what is in the dolphin's stomach," I sheepishly replied. Though I was on the receiving end of a bewildered look, I had my subject and was off to perform the postmortem analysis. Once the stomach's cavity was sliced open, two partially digested flying-fish, three crabs, one sargassum fish, and a mollusk shell spilled out on the table. Quite a feast, I thought, and then my eyes drifted over to a small, round object still lodged within the stomach. This bull dolphin, in his rush to feed, had inhaled a beer bottle cap — unfortunately part of the increasing jetsam found within the Gulf Stream.

One of the intriguing aspects of trolling the rich waters of the Gulf Stream is the angler never knows for sure which particular game fish may strike a lure or bait. But the line-burning run of a hooked wahoo is undeniable. The wahoo is capable of attaining speeds of fifty miles per hour in short bursts, solidifying its reputation as one of the fastest fish in the sea. Any angler who has hooked a large wahoo and watched it sizzle a hundred yards of line off the reel in a few seconds will attest to this.

The wahoo was originally found to be in great abundance near the Hawaiian island of Oahu, which according to some accounts was often spelled "Wahoo," possibly explaining the origin of this fish's name. On a more formal basis, Georges Cuvier, a noted anatomist and paleontologist, first described the wahoo, a member of the tuna/mackerel family, in 1831. Originally named *Cybium solandri*, Cuvier later changed it to *Acanthocybium solandri*. The genus name is derived from the Greek *akantha*, meaning "thorn," and *kybion*, "scombrid." The first word is an apt description of a wahoo's most distinctive feature: dozens of large, triangular, finely serrated teeth. These teeth are put to good use since wahoo are almost totally carnivorous. Research performed in North Carolina demonstrated that fish, including herring, jack, mackerel, and flying fish, accounted for more than 97 percent of their diet. Like a lone wolf, wahoo are often solitary hunters or travel in very small, loosely associated groups, never schooling like dolphins. Savvy anglers who are knowledgeable about the predatory habits and food preferences of wahoo will often troll fishlike lures in straight lines at relatively high speeds, above twelve knots. If these surface lures do not elicit a strike, anglers may employ downriggers and planers, which position the lures farther down in the water column to target deep-swimming wahoo. Regardless of the methodology, a wire leader, inserted between the hook and the main line, is good insurance against the formidable dental work of the wahoo. Nothing frustrates a crew and angler alike more than a cut-off allowing their prize to swim away with an expensive lure firmly planted in its mouth.

Wahoo (copyright iStockphoto.com/Chuck Babbitt)

Many an angler can attest to the fierce nature of the wahoo. A wahoo hoisted on board will violently thrash about, wreaking havoc where and when it can — and remove a few digits from a careless mate in the process. But the wahoo is a prime target for many anglers, who appreciate the more delicate texture and flavor of its white flesh compared to the stronger flavored, blood-red flesh of the long distance cruisers, such as bluefin tuna. The Hawaiian word for this fish is *ono* — literally, "good to eat" — which probably best sums up its popularity as a seafood item.

The last of the "meat" fish trinity (those caught mainly for human consumption) is the yellowfin tuna. Pierre Bonnaterre, a natural history professor, first described this species in 1788, originally naming it *Scomber albacares*. Over the years, the fish appeared under a variety of names (more than two dozen) until 1953, when it was renamed *Thunnus albacares* — from the Greek *thynnos*, meaning "tunna."

Because of the yellowfins' strong schooling tendency, it is not uncommon to hook three to five fish almost simultaneously. These multiple hookups are the norm rather than the exception and require all the skill that the angler and mate can muster to ensure a successful catch. Many anglers rank the yellowfin as one of strongest fighting fish in the ocean. With the call to action of "fish on," the anglers must steel themselves for a long, tough battle with this pelagic adversary. While wahoo are the greyhounds of the Gulf

Stream, yellowfin tuna are the bulldogs — diving deep and slowly swimming in circles beneath the boat. To the angler it may feel like fishing's version of Dante's levels of hell when a determined yellowfin seeks its freedom. Since these fish can attain weights of more than a hundred pounds, the angler must exhibit the proper technique in fighting such a large fish — make a wrong move, and the fish is gone. The trick is to retrieve as much line as possible when the fish is on the inward leg of the circle and minimize yielding any line as the fish completes the outward portion. Armed with a stout rod, the angler settles into an efficient routine of "pumping the rod," pulling it back to gain line and quickly lowering it to wind the line on the reel. Captains will often aid the angler who is engaged in a tug-of-war with a large fish by "backing down the boat," repositioning it so that the angler is not pulling at an acute angle to the fish.

Though the diet of yellowfin tuna includes fish, cephalopods, and crustaceans, which they forage rather indiscriminately, they can, like many game fish species, develop "lock jaw." When the boat hasn't landed a fish in quite some time, you're sure to hear, "Okay, who brought the bananas?" How did this fruit come to be thought of as bad luck on a boat? Superstition has long been part of maritime tradition, and probably no other lore holds sway with captains and mates more than that having these crescent-shaped, yellow fruits aboard will jinx the vessel. There are many stories as to why bananas gained their bad reputation, and many of them have their origin in the days of wooden sailing ships of the early seventeenth century. As the Spaniards traded throughout the Caribbean, their vessels would stop at tropical islands to gather provisions. Often the crew would purchase wooden crates of bananas and bring them on board the ship. Sailing north out of Havana to pick up the Gulf Stream in the Straits of Florida, many hazards might befall these vessels, resulting in shipwrecked or scuttled ships. The only remains of these sinking ships might be clumps of floating bananas, leading witnesses to conclude that hauling this cargo was unlucky. As captains circulated the rumor that bananas were bad luck, the crew and passengers may have been quite willing to forgo this provision because of the inherent danger of these open-water crossings. Another variation of the above story is that these crates would have all manner of unwanted pests, such as bugs, spiders, and vermin, which would ultimately find their way into the captain's quarters. Attempting to keep these critters off the ship, the captain might spin a tale of bananas being bad luck. And yet, over the years, some bold individuals have scoffed at this superstition and developed banana-shaped lures, which they claim mimic a wounded baitfish. Hype or truth? What we do know is that none of these lures has become part of the tackle arsenal of the offshore charter fleet.

Bananas aside, one of the most difficult Gulf Stream game fish to catch is the billfish. Three species, the sailfish, white marlin, and blue marlin, are all part of the same biological family, *Istiophoridae*, and are readily recognizable by their long, spearlike beaks. Though there are differences in size and anatomy (sailfish have a large dorsal fin from which they derive their name), all billfish have adapted remarkably well to their pelagic existence. These high-speed predators of fish, squid, and crustaceans exhibit an upbeat lifestyle, as reflected in their coloration, body form (fusiform), and physiology (warm-bodied). The patterning and coloration of billfish provide them with the perfect camouflage for their home in the sea. The body is dark blue dorsally, allowing the billfish to blend in with the darkness of the abyss when viewed from above, but shades to a silvery white ventrally, making it difficult to discern from the sunlit background when seen from below. The coloration of a marlin may change markedly, depending on its level of excitement. It may become "lit up" — the blue bars along its flanks dramatically increasing in intensity and contrast. While anglers marvel at this explosion of color, studies have shown the brilliant blue stripes of some billfish contain a strong ultraviolet component that is invisible to humans and billfish. If a billfish cannot see the ultraviolet component of its own body or of its billfish cousins, why does the fish display it as a very prominent part of its light show? One hypothesis is that this conspicuous coloration may have a disruptive effect on a school of prey: confusing it, leading to a breakdown of the school's tight-knit structure. Divers have observed on a number of occasions a sailfish slowly circling a school of prey and then suddenly darting in to pick off a dazed member.

For years, scientists have puzzled over the purpose of a billfish's formidable bill. Is it used as a weapon to slash through and stun a school of prey? Analysis of the stomach contents of billfish yields little evidence of any slashed fish, leading to the more widely accepted conclusion that the bill may be an adaptation for speed, similar to the pointed nose of a missile.

Of all the billfish species, many anglers rank the blue marlin as the top

Gulf Stream sailfish (copyright iStockphoto.com/Rafael Ramirez Lee)

game fish in the sea. With a rare combination of size (the all-tackle record for the Atlantic Ocean is 1,402 pounds), power, and jumping ability, the blue marlin is the ultimate test for any angler. And so it was for Santiago, the protagonist in Hemingway's novella, who is amazed by the size of his hooked marlin, "two feet longer than the skiff." He realizes that the marlin could easily destroy the boat if he wanted to, pleading, "Thank God, they are not as intelligent as we who kill them, although they are more noble and able." But upon seeing the fish for the first time, Santiago becomes obsessed with the idea of proving himself a worthy slayer of such a grand beast. Present-day anglers can definitely identify with Santiago's obsession as they pursue this marvelous species. But this was not always the case. Tom Carlson writes, in his book *Hatteras Blues*, that many of the old-timers "didn't want to fool with these things — too big, too scary."

Bernard Lacépède first described the blue marlin, *Makaira nigricans*, in 1802. The genus name *Makaira* is derived from the Latin *machaera*, which means "sword." But the taxonomic status of the blue marlin is open to debate. Some within the scientific community consider the blue marlin a species common to all the tropical and semitropical waters of the world's oceans, while others, based upon morphological differences, claim the blue marlin of the Pacific and Indian Oceans are separate species.

The blue marlin is unequivocally at the top of the Gulf Stream food chain. As the apex predator, the marlin preys upon other pelagic fish, such as mackerel, dolphin, and tuna, with considerable frequency and ease. (A forty-pound dolphin makes a nice snack for a blue marlin.) Based upon this dietary preference, some marlin anglers fervently believe big baits get big fish. While a mate may expertly rig a whole mackerel or even a small tuna to test this hypothesis, a marlin's interest in these baits may be more than satisfying its gastronomical desires. All higher-level organisms, including the marlin, use their senses to interpret the environment around them, focus in on specific objects or tasks, and filter out extraneous noise. What mental process might allow the marlin to focus its attention on the larger offering (target) and ignore the smaller lures or bait (distractions)? If the target is markedly different in appearance from the distractions, the target is said to "pop out" in the mind of the receiver of this signal. Cognitive research has shown that if the target is extremely distinctive, the number of different distractions has little or no bearing on the selection of the target. While this premise is likely to generate considerable debate within the offshore angling fraternity, fly-fishing anglers have consistently used "attractor patterns" of artificial flies with considerable success. Gaudy-colored salmon flies, for example, clearly have no natural counterparts, which are generally drab and muted in coloration. But these flies elicit a feeding response in fish even when there are many natural food items (distractions).

Regardless of the blue marlin's choice of food, the sight of one charging into the trolling spread at full speed — flanks lit up, lunate tail slicing through the water — is probably the ultimate adrenalin rush for the angler. The thrill of hooking a marlin can be quite intoxicating when the fish makes a series of acrobatic jumps, sustained runs, and deep soundings — all in an attempt to secure its freedom. But critical to the survival of the marlin is how quickly and adeptly the fish is overcome and brought in to the boat. Essentially, a short fight and a quick release markedly improve the fish's chance of survival. After fly anglers in Michigan instituted the practice of catch-and-release during the 1950s as a means of reducing the need for trout stream stocking, many conservation-minded anglers adapted this technique to the marine environment. Because of the economic value of these fish to the Gulf Stream fisheries and concern about the decreasing population of the species from commercial long-line fishing, most anglers release marlin to fight another day. As Lee Wulff, a noted fly fisherman and angling author, put it, "A sport fish is too valuable a fish to be caught just once." It appears this mantra has taken on wide appeal in the offshore angling community. Though Hemingway's Gulf Stream fishing exploits were immortalized in photographs of him beside a dead marlin, recognition for today's angler is more subtle: a marlin flag aboard the boat is flown upside down to signal a marlin release. The ultimate accomplishment in offshore angling is the rare "grand slam," the release of a sailfish, white marlin, and blue marlin all in the same day.

The degree of fight in a fish depends upon the physiology of its muscles and its metabolic responses to increased stress. As the fish struggles to regain its freedom, the oxygen in its muscles decreases, and metabolic wastes (lactic acid) begin to accumulate in the muscle tissues, leading to extreme muscle fatigue and a fish that is no longer up to the battle. If the fish is unable to eliminate this lactic acid buildup in a timely manner, it will succumb to its weakened state. Fighting a fish for an extended period of time, especially in warm water, reduces its chance of survival.

The survival of an exhausted marlin depends upon the coordinated action of the mate and helmsman to revive the fish before releasing it. After the mate brings the fish to the boat, he holds it in the direction the boat is moving to force oxygen-laden water over its gills. Once the fish has revived — a twitching of its bill or its tail will signal that — it's time to let it go. Though the number of marlin that survive this ordeal is unknown, a marlin that is boated and unceremoniously dropped on the deck has no chance of survival.

Survival of the game fish in the trackless and featureless ocean depends upon efficient predation. Sight-dependent predators are ideally suited to the clear waters of the Gulf Stream. Their well-developed visual system strongly suggests that vision is an important sense for the detection and interception of prey and lures. As consummate predators, marlin rely strongly on visual perception, with almost 30 percent of their brain devoted to analyzing the visual signals coming from the eye. The concept of perception is a bit elusive. If an animal perceives something, that implies the animal not only detects the presence of an object but also identifies it as fitting into a particular category. For a wahoo to perceive a fleeing mackerel, it must detect the object in its field of vision and recognize that object as a food item, rather than, let's say, the "banana" lure.

The ability of any animal to see depends on two components: a means of collecting the ambient light and a method of forming an image. While humans rely on a cornea and lens to focus an image on the retina, the anatomical setup of fish does not include a cornea because it would be essentially useless underwater. Pelagic fish must rely solely on a lens to focus the light into a clear image. The spherical-shaped lens is moved in and out, like the zoom lens of a camera, in contrast to stretching the way the lens of a human eye does, to generate a sharp image.

A fundamental measure of visual capability is sensitivity: the minimum amount of light that enters the eye resulting in a behavioral response in the organism. Unlike terrestrial vertebrates, fish cannot dilate or contract their pupils in response to changing light conditions. They have no anatomical means to control the amount of light entering the eye. But pelagic species differ in their ability to detect objects in low light intensities. A dolphin, for example, forages within the upper sunlit layers of the Gulf Stream and mainly during daylight hours. In contrast, the optical sensitivity of a marlin's larger eye is higher because of the large number of light receptors it contains. The marlin's eye is specifically adapted to cope with low light levels encountered during a dive, thereby increasing its feeding range. These large, optically sensitive eyes may also aid the marlin in seeing while it swims at high speed in search of prey.

Given adequate light, what is the visual acuity of a pelagic fish? What is its ability to resolve the smallest detail of an object? Or from the predator's perspective, what is its ability to discriminate between food and nonfood items? Though numerous methods exist to measure visual acuity, the most common, by far, is to determine the minimum angular size of standardized letters that can be identified — "reading the eye chart." If "normal" human vision is 20/20, then the visual acuity of dogs is approximately 20/50 to 20/70, and in marlin it is 20/90, allowing them to resolve a four-inch object from a distance of 150 feet under optimal light conditions. While this acuity is typical of most fish, it is surprising in billfish, given their large eyes.

Do pelagic species, like tuna and billfish, distinguish colors? This question has not only generated interest within the scientific community but has captivated offshore anglers, who fill their tackle boxes full of colorful lures to attract Gulf Stream game fish. While direct color-discrimination experiments are virtually impossible to undertake with large fish in the open ocean, anatomical evidence strongly supports the existence of color receptors, known as cones, in the eyes of these animals. Human eyes are sensitive to a rainbow of colors (violet to red), but the most abundant pigments found in the photoreceptors of tuna and marlin are blue-green. These pigments are perfectly matched to the prevailing light that filters down into the clear ocean water. Since the color red is filtered out rapidly with depth, these species' visual systems have evolved to optimize color sensitivity at deeper diving depths. While a bright red lure may generate significant appeal as the "hot" lure of the day to all those on board the boat, it may elicit little interest from a tuna that is essentially color-blind in the red end of the light spectrum.

If light conditions are far from optimal or fish are not picking up visual cues, they may rely on mechanical perception, sensitivity to vibrations in the water, for detecting prey. All animals produce waves when moving through water that can be detected as sound. Since sound travels quickly in water — five times faster than in air — water is like an early warning system for the approach of prey or predator. Many fish can sense and interpret the sound from disturbances in the water by using both their ears and lateral line organ.

While having no visible external ear, most fish have an inner ear that consists of two parts: an upper section for balance and equilibrium and a lower section composed of otoliths (small particles of calcium carbonate) that vibrate with sound. The fish's lateral line provides it with a cross between hearing and touch. The organ responsible for this is the neuromast, a cluster of very tiny, hairlike cells. All fish possess free neuromasts, which come in direct contact with the water and are arranged linearly to form the fish's lateral line.

The important thing for the predator is that these sensory systems pick up "good" vibrations that the fish can process and interpret as an available meal. A frightened prey that is attempting to flee produces a complicated series of swirling motions, which are presumably detectable by a predator several yards removed. The game fish may detect and evaluate the origin of such vibrations by swimming through them or by perceiving them as auditory cues as they sweep past in the flow of the Gulf Stream. A whole cottage industry has developed around the premise that game fish are attracted to particular sounds that are produced by trolled lures. With such names as Big Chugger, Plasmic Jet, and Super Smoker, these hydrodynamically designed lures claim to stir up the water surface by creating an enormous bubble trail in their wake. Manufacturers tout each new innovation in lure technology, as evidenced by this claim: "Armed with government research and the help of an aerospace scientist with twenty years of experience in designing microwave and millimeter wave electronics, the folks at Sevenstrand have developed a teaser unlike any other. Each features a tiny device that replicates distressed baitfish sound frequencies."

Even the hum of a diesel engine, the pressure waves from the bow of the boat, or the pounding of the boat's hull through the water may generate distinctive vibrations that, like a dinner bell, call the fish to eat. Rising from the depths, billfish are lured to the surface to satisfy their curiosity about the nature and source of the sound. Anglers commonly refer to this vertical migration as "raising" billfish.

But raising a fish is not catching a fish, nor is it even hooking the fish. In some cases, the proportion of billfish that are hooked is relatively low compared to those lured to the surface. Does the fish spurn the offering because of lack of interest or hunger? Does it miss the target (bait/lure) in spite of the visual and/or auditory signals from the target? Though almost every blue-water angler can attest that there are times when a marlin is neither curious nor hungry, why a marlin misses a feeding opportunity, though stimulated by environmental cues, is not so clear. Recognizing that billfish, for whatever reason, sometimes miss their food offering, most seasoned anglers are pleased if their efforts lead to a success rate of one or two marlin caught for every dozen enticed to the surface. The Gulf Stream is reluctant to relinquish its prize inhabitant.

When all else fails, including trolling noisy lures, bait, and various teasers, in catching these denizens of the Gulf Stream, what is a frustrated captain to do? He can't go wrong taking the sage advice of the admiral himself, Christopher Columbus: follow the birds. On October 7, 1492, Columbus's fleet had been almost a month at sea and still had not sighted land. Almost mutinous, the crew of Columbus's small fleet desperately wanted him to abandon his quest and turn back. Toward evening, Columbus observed a large flock of birds flying from the north to the southwest. He interpreted their flight pattern to mean either the birds were going to land to nest for the night or were migrating to warmer lands in response to the approaching winter. Based upon this observation and the knowledge that the Portuguese had discovered most of their island possessions by being cognizant of the behavior of birds, Columbus altered his course from west to the west-southwest. Determined to follow his instincts, Columbus doggedly maintained this course, and his persistence paid off; he sighted land four days later.

Near the North Carolina coast, the interplay of the warm Gulf Stream with the cold water on its western edge supports a remarkable variety of seabirds, including shearwaters, terns, and the black-capped petrel (*Pterodroma hasitata*) — one of the signature species of the Gulf Stream — which feed upon the baitfish that attract the dolphin, wahoo, and tuna. Persistent and observant captains are constantly on the lookout for large flocks of birds congregating over a bait pod that has been driven to the surface by hungry tuna. The water "boils" with activity as fish slash through the bait and screeching birds swoop down to pick at the morsels. Once a savvy captain locates the flock and ascertains its direction of travel, he will motor ahead of the flock to intercept its path and, hopefully, the accompanying fish. Trolling through this frenzied triangle of birds, bait, and game fish often elicits multiple strikes and hookups from the latter.

While modern-day mariners generally welcome the sight of birds as a good omen, the brown booby (*Sula leucogaster*) generally does not represent good news for the angling community. Boobies were apparently given their common name because of their fateful mistake of allowing sailors to approach and kill them for food. These "slow learners" are generally content to sit on the water or near a patch of sargassum weed, hoping that a meal will simply come their way. While the propensity of these birds to avoid foraging offers no help to the angler looking for signs of life in the pelagic environment, it can turn to utter frustration when these birds repeatedly attack his trolled lures and bait. Too lazy to secure their own meal, the brown boobies sometimes resort to outright stealing — at least, that's how the angler sees it. Just as it was lucky for Columbus that he was familiar with birds, such as the long-tailed tropic bird or the red-billed tern, during his journey, a captain who is knowledgeable about the Gulf Stream's avian residents can often turn an unfruitful fishing trip into a lucky one.

While luck certainly enters into the fishing equation, history reveals that early mariners needed more than luck to successfully navigate the vast expanse of the open Atlantic Ocean. Hard-won knowledge of ocean processes and intimate familiarity with the winds and currents would contribute to the success of the explorers, discoverers, and settlers of the New World.

PART 3 SAILING THE ATLANTIC

7 EXPLORATION AND DISCOVERY

THE VIKINGS WERE THE FIRST Europeans to traverse the North Atlantic between 900 and 1200 A.D., but centuries would pass before other European nations would sail these waters. Long ocean voyages became a part of history and lore, read and spoken about, but not attempted. What precipitated the decline in the number of transatlantic voyages? Though historical documents provide little information to answer this question, we can speculate as to a possible contributing factor.

As the age of the Vikings was coming to a close, a cool-down of the northern hemisphere was just beginning. One of the longest climate records available to researchers lists the frequency of sea ice off the Icelandic coast. Because of their strong dependence on fishing, Icelanders kept meticulous records of the times when ice blocked harbors and made going to sea impossible. This time series reveals an interesting pattern: essentially ice-free years between 1000 and 1200, and sea ice present an average of five weeks per year between 1200 and 1400. Even the most adventurous mariner may have thought twice about undertaking a long voyage during this period of ice-choked seaways.

Changes in climate are inevitable: cooling trends followed by prolonged warming spells, or drought conditions preceded by rainy periods. Another look at the sea-ice record shows an absence of ice starting around 1400 and stretching to about 1500, which marks the beginning of the Little Ice Age. Had this short and relatively weak warming interval provided another window of opportunity for more seafaring nations?

While historians disagree about whether early fifteenth-century Europeans were aware of even rudimentary features of the Atlantic circulation, the consensus is that they did at least suspect the existence of ocean currents. The recovery of strange flotsam at Porto Santo in the Madeira Islands and unknown wood types near the Cape Verde Islands appeared to confirm their belief in current flow in the open ocean. Columbus was particularly fascinated with a Caribbean seed, the sea heart, which routinely drifted ashore on the beaches of the Azores. These seeds, named for their resemblance to large wooden hearts, are said to have provided the inspiration to Columbus and convinced him that new lands lay to the west. Even today, the sea heart is called *fava de Colón*, or the "Columbus bean," by residents of the Azores. The belief in sea hearts as good omens for seafarers would carry on long after Columbus. English sailors would carry sea hearts on board their ships, rationalizing that if these seeds could stay adrift a year or longer to reach European shores, then they could protect their owners on their long and perilous journeys.

Though foreign objects that washed ashore may have tantalized Europeans about the possibility of transoceanic flow, economic necessity drove them to become intimately familiar with the subtle interplay between wind and water. Seafarers did not undertake exploration for its own sake; any voyage had to be firmly based on achieving a specific material goal. Without the prospects of obtaining wealth or territory, there would be little incentive on the part of European royalty to fund an ocean expedition.

The first step in European ocean exploration came with the assault by the Portuguese on the Moorish port of Ceuta in 1415. Ceuta had been the hub of a centuries-old Saharan caravan route that provided a source of gold, rumored to have originated in the headwaters of the rivers Niger and Volta. The temptation of this wealth was too great for the Portuguese to resist; it became the impetus for discovering a sea route down the West African coast that would bypass the caravan route, presently controlled by their enemies, the Moors.

Perhaps the most influential individual during this period of exploration was Prince Henry, who had led the charge against the Muslim forces at Ceuta. Three years after his military service, Prince Henry founded a center for the study of marine science and navigation at Sagres, Portugal. With the financial backing of his father, King John I, he worked tirelessly to bring together at the institute Europe's finest geographers, cartographers, astronomers, and mathematicians. Under his leadership, expeditions sailed

Caravel with lateen sails (copyright T.W./Shutterstock.com)

down the West African coast that would start a half-century of unprecedented maritime discovery. Though he personally never sailed on any of these voyages, he became known as Prince Henry the Navigator.

Both the southward-flowing Canary Current and the northeast trade winds would aid discovery ships sailing from Portuguese ports to equatorial Africa. But a discovery ship was not a cargo-hauling vessel since it had to be prepared to go long distances in uncharted waters and return. An exploring ship was of little value unless it could bring back its findings — its important cargo was news. The difficulty was that most of the ships of this period sailed poorly against the wind. Rigged with square sails, they were at their best with the wind at their stern. In order to solve the problem of sailing into the wind on their return leg, Prince Henry's shipbuilders produced the caravel, which combined Arab and European shipbuilding techniques. This small, highly maneuverable ship was able to beat against the wind because of its lateen (triangular) sails. The caravel was to become the standard-bearer of many explorers' voyages.

Between 1424 and 1434, Prince Henry sent fifteen expeditions south along the African coast; each returned without completing its journey. Portuguese captains were reluctant to venture around the dreaded Cape Bojador on the western bulge of Africa. Mariners told stories of hideous sea monsters devouring any ship that sailed south of this promontory into the "Sea of Darkness." Even less superstitious Portuguese feared the cape was a point of no return; once it was passed, they believed, it would be impossible for a ship to return north against the prevailing winds. But Henry was determined this invisible barrier would be crossed. He was driven as much by a crusading fervor — the need to save the souls of the infidels in the unknown lands — as by the need to expand the maritime interests of Portugal, a rising economic power in Europe.

For the task, Prince Henry chose Captain Gil Eanes, who had served Henry from boyhood as a squire. Urged by Henry to "make the voyage from which, by the grace of God, you cannot fail to derive honor and profit," Gil Eanes would attempt to round the cape in 1434 on what would be his second attempt. Terrified but determined to redeem himself for his past failure, he and his crew set sail in a small fishing boat. Aided by favorable winds, they soon reached the Canary Islands to resupply their vessel. The Canary Islands are just north of Cape Bojador, and it was at this point that Gil Eanes debated his decision to continue. His confidence was wavering, but he then made a bold decision. Instead of sailing near the cape, he would sail far to the west of it; none of the crew would see this dreaded land. In retrospect, it was a wise choice. The Arabs called Cape Bojador *Abu khater*, meaning "father of danger," and, indeed, the cape is known for its numerous shipwrecks. The combination of reefs and shoals (with less than six feet of water), which extend more than three miles from shore, led to many ships running aground. Gil Eanes skirted the cape unscathed and joyfully made the announcement to his relieved crew. Emboldened by their good fortune, they sailed further into the fabled Sea of Darkness. Instead of the horrors of legend, they found themselves engulfed by the hot, dry Saharan wind, known as the harmattan. The blinding, dusty wind forced them to land on the desert coast. As proof he had successfully carried out his charge, Gil Eanes collected and brought back to Prince Henry a hardy plant that came to be known among the Portuguese as "St. Mary's roses."

In the next few years, most of western Africa would be visited by the Portuguese, well before Columbus's own voyages of discovery. This opening of new sea-lanes would culminate with the long-anticipated rounding of Africa in 1486 by Bartholomeu Diaz. Unfortunately, few records of these early voyages have survived, creating a void about what was learned of ocean currents. While some of these documents physically deteriorated over time, more likely the Portuguese royalty, who financed these voyages, cast a cloak of secrecy over them to protect their economic interests. This practice of closely guarding any scientific information that would have an economic impact long concealed from public view the early discoveries made during seafaring expeditions, such as those to the New World.

As new discoveries stretched the bounds of human knowledge of this planet, fifteenth-century cartographers assimilated and transformed this information into usable forms. Martin Behaim in Nuremberg constructed one of the first globes delineating the distribution of land and water on earth in 1492. It showed a vast expanse of water called the Ocean Sea, which separated Europe and Africa from Asia. This globe and the 1490 world map by Henricus Martellus depict an archipelago off the Asiatic coast. Buoyed by these views of the known world of his time and the lure of the riches from the East Indies, Christopher Columbus believed a westward voyage from Europe would reach the Indies. Contrary to the popular view of his time that the earth was flat, but like most better-educated people, Columbus was convinced the earth was a sphere, but he estimated its circumference to be about three-quarters of its actual size. Based upon this misinformation, he erroneously calculated that the distance from the Canary Islands to Japan was only twenty-seven hundred miles and that he would encounter these islands along the way. Since the actual distance is more than ten thousand miles farther, Columbus would probably never have undertaken his voyage of discovery if he had been aware of the actual distance. But he did go. And as every grade school history book recounts, Columbus did not discover a westward passage to the Orient, since the body of water he sailed upon separates Africa and Europe in the east from the Americas in the west, not the East Indies. The Ocean Sea of Columbus's period is now known as the Atlantic Ocean, named for Atlas of Greek mythology.

While Columbus was grossly mistaken about the continents, he did know the sea. And to know the sea in his time especially meant knowing the winds. He acquired most of that knowledge from his training by

Portuguese navigators and pilots, years before his epic voyage. Sailing with Portuguese expeditions down the African coast and touching at the Cape Verde and Canary Islands, Columbus must have become fully acquainted with the trade winds. The experiences of these early years led directly to the genesis of his plan to set his course southward from Spain to the Canaries, instead of sailing due west against the prevailing westerlies, as other, failed expeditions had previously attempted. At the Canaries, he turned due west, exploiting, possibly unknowingly, the dual advantage of the trade winds and the North Equatorial Current. Even a cursory look at Columbus's log shows he was unaware that these winds, never mind the ocean current, would carry him across the Atlantic. On the other hand, some historians attribute his success to intuition on his part: by turning westward at 28° north latitude, he would eventually encounter the islands of his quest. His underlying premise was simple — sail west to reach the East. But one gains some additional insight into the possible sea knowledge of Columbus by looking at the design of his three ships. The *Santa Maria*, the flagship of this fleet, had two square sails on the foremast and mainmast but only one lateen sail on the mizzenmast at the rear. Outfitted in this manner, this big and lumbering ship could take maximum advantage of the trade winds. While both the *Pinta* and the *Niña* were smaller caravel ships, the *Pinta* most likely carried square sails, like those of the *Santa Maria*. In addition, though the *Niña* left Spain with lateen sails on all masts, Columbus ordered she be refitted in the Canary Islands with square sails on both fore- and mainmasts to take advantage of the favorable winds.

Once underway from the Canary Islands, the ships benefited from calm seas and the steady push of winds against their fully rigged masts. But as the passing days turned into weeks, and still no land in sight, the crew became fearful. They were, as most mariners of their day, a suspicious lot, prone to spinning tales about sea monsters, a flat earth, and the boiling water of the scorching tropics. The trip was becoming long, longer than Columbus had anticipated. They had already sailed further than any known European. In order to placate his crew's apprehensions, Columbus kept two sets of logs: one depicting the true distance sailed each day, and the other showing a lesser distance. Columbus never revealed the first log to his crew, but he did make known the contents of the latter — essentially, a fabrication of the true distance they had traveled. He hoped his deception would buy him more time to reach his destination.

When Columbus's fleet was ten days out from the Canary Islands, they thought they had reached the point on Columbus's map where the island of Antilia should be located. Samuel Eliot Morison, considered the chief biographer of Columbus, relates that Columbus, excited by this possibility, ordered the sounding lead be readied to determine the depth. Two lines were spliced together, totaling twelve hundred feet, and tossed over the side. But the lead never found the bottom, since at their location the actual ocean depth was about thirteen thousand feet. Columbus had grossly underestimated not only the distance to the Indies but also the depth of the abyss.

His mistakes were starting to wear on the crew, who found the daily grind of shipboard chores increasingly intolerable; they were reaching their breaking point. The tension among the crew heightened even more when they found themselves adrift in the Sargasso Sea during the last week of September, becalmed by the weak winds and no currents. Columbus immediately recognized the gravity of their situation. With the distinct possibility of running out of potable water, Columbus and his crew faced the ominous prospect of dying from thirst. They could not drink seawater because of the salt: since the kidneys can only produce urine that is less salty than seawater, they would have to urinate more water than they drank — the result being a slow death by dehydration.

The despondent men pondered their fate, wondering whether they would ever see their homeland again. Where were the favorable winds that Columbus promised? After two agonizing days in the Sargasso Sea, a freshening breeze billowed the sails of the ships and lifted the spirits of the men. They had drifted back into the realm of the trade winds. The westward trek to the Orient continued.

Columbus never wavered in his resolve to find the riches of the Indies. Though he had private quarters in which to rest, he stayed on deck throughout much of the day and night, scanning the horizon for any sign of land. But by October 10, the crew was on the verge of mutiny. To defuse this volatile situation, Columbus promised that if land were not sighted in two days, they would return home. The next day, to the relief of all on board, they caught sight of land.

Aided by the relatively favorable prevailing winds and currents, Columbus's ships had sailed thirty-three days from the Canary Islands, averaging slightly more than one hundred miles per day at a speed of approximately four knots. While that was slow compared to the speed of modern racing vessels, the sailboat skippers of today would be hard-pressed to find a better westward route across the Atlantic than along the southern half of the North Atlantic gyre.

But where exactly was Columbus's landfall in the New World? This question has led to considerable debate among historians, scientists, and maritime enthusiasts over the last five hundred years. In an attempt to answer this question, two oceanographers, Philip Richardson and Roger Goldsmith, from Woods Hole Oceanographic Institution, reconstructed the wind and current conditions that Columbus reported in the log of his voyage. From this information, they argued that Columbus's original landfall was the Bahamian island of San Salvador. As far back as 1942, Samuel Eliot Morison in *Admiral of the Ocean Sea* had championed San Salvador as Columbus's landing. But a new challenge arose in 1986 when Luis Marden, writing in *National Geographic* magazine, named Samana Cay, a tiny island in the Bahamas, as the landfall. By the 1990s, this renewed interest in the subject ushered in several new landfall theories that now include nine widely diverse islands scattered over three hundred miles throughout the Bahamas.

After his landing, Columbus steadfastly continued to believe he had set foot on the fabled lands of the East Indies. He tirelessly searched other islands, including Cuba and Hispaniola, for the magnificent cities Marco Polo had described. All his efforts, including two subsequent voyages to the New World, proved to be in vain. But despite evidence to the contrary (such as Amerigo Vespucci's 1501 voyage to South America), Columbus resolutely maintained, even to his death in 1506, that he had found a new route to Asia.

Though Columbus may have turned a deaf ear to the discoveries of other explorers, he was a master at navigation, a skill honed by his years of sea voyages. In 1493, on his second voyage to the New World, Columbus planted the Spanish flag on the island of Dominica in the Lesser Antilles. Covering twenty-five hundred miles in little more than three weeks, Columbus had sailed the shortest possible route from the Canary Islands to the New World. If he had miscalculated, setting his course just five degrees higher, he would have missed the Lesser Antilles entirely and might have had to sail an additional five hundred miles to reach Cuba.

Arne Molander and James Norris, longtime Columbus admirers, hypothesize that the discoveries on Columbus's second voyage were not a matter of luck at all, but the result of his ability to compile and assimilate bits of information. On his first voyage, Columbus had anchored in Samaná Bay. The friendly natives he encountered regaled Columbus with stories about a group of islands that lay to the east of Hispaniola. Were these the Asiatic islands he was seeking? How could he even be sure these islands existed? A clue would come from the brown seaweed that accumulated in Samaná Bay. Since Columbus was familiar with the pelagic form of sargassum, he immediately recognized this seaweed as being different in appearance, having holdfasts that anchor it to the bottom. But where had it come from? Columbus came to the conclusion that this benthic (from the ocean bottom) form of sargassum had been torn free by storms and carried by currents to its final resting place. He must have known of the currents in the Caribbean because he commented frequently on their existence. In his log he writes, "When I left the Dragon's mouth, I found the sea ran so strangely to the westward that between the hour of Mass, when I weighed anchor, and hour of Complines, I made sixty-five leagues of four miles each with gentle winds." Peter Martyr, the official historian of Spain at this time, chronicled Columbus's encounter with a current on the Honduran coast: "He left in writing that sailing from the Island of Guanassa toward the east he found the course of waters so vehement and furious against the fore part of his ship that he could at no time touch the ground with his sounding plummet, but that the contrary violence of the waters would bear it up from the bottom. He affirmeth also that he could never in one day with a good wynde wynn one mile of the course of the waters." The Antilles Current, which flows northwestward east of the Antilles, appears to be the most likely conveyor of the seaweed — a current Columbus most likely sailed on.

Though Columbus on his three voyages extensively explored the Caribbean basin, he never traveled far enough north to directly encounter the Gulf Stream. Being driven as he was to find the Indies, he must have considered sailing northward at some time. What deterred him from expanding his search? Historians have reconstructed from his logbook a possible explanation, one based on a weather event. As Columbus approached the Caribbean basin, a cold front pushing south off the Florida peninsula altered his fleet's course. Northerly winds behind the front forced them to sail along a more southerly course. Had the front not impacted the region, the trade winds would have steered the fleet into the Gulf Stream and then up the coast, where most likely they would have made landfall along the eastern seaboard.

By the fifteenth and sixteenth centuries, the door to the seas began to shut as the maritime superpowers of Portugal and Spain began making preposterous claims to vast areas of the world's oceans. Emboldened by Columbus's voyage of discovery, Spain petitioned Pope Alexander VI (a Spaniard) to issue a papal bull that ostensibly awarded almost all of the New World to Spain. It forbade any nation to cross, for whatever reason, a line drawn at 35° west longitude. Spain would essentially control all access to the waters of the western Atlantic and the Gulf of Mexico. Along with Portugal's claim to exclusive navigational rights over the rest of the Atlantic and the Indian Ocean, this order ushered in the dark period of closed seas, or *mare clausum*.

Five years after Columbus's discovery of the West Indies, another Italian explorer, Giovanni Caboto, sailed from Bristol, England, intent on finding a westerly sea route to Asia. Like Columbus, Caboto made another country, England, rather than his homeland, Italy, his base of operations, and thus the English-speaking world recognizes him by his English name, John Cabot. He surely must have known about Columbus's southern route to the Indies, but Cabot was strongly opposed to following Columbus's sea-lane. He petitioned King Henry VII of England to support his venture because he believed, incorrectly, that the spices from the Orient originated in northern Asia. Cabot thought that England, located at relatively high latitude compared to the Mediterranean countries of Europe, would be a natural departure point from which to sail north rather than west. Furthermore, King Henry had granted him a patent to "sayle to all partes, countreys of the East, of the West, and of the North" but strictly prohibited his sailing southward into waters controlled by the Spanish.

Detailed information about Cabot's voyage is simply not available from modern archives. If he kept a log or made maps of his journey, they have

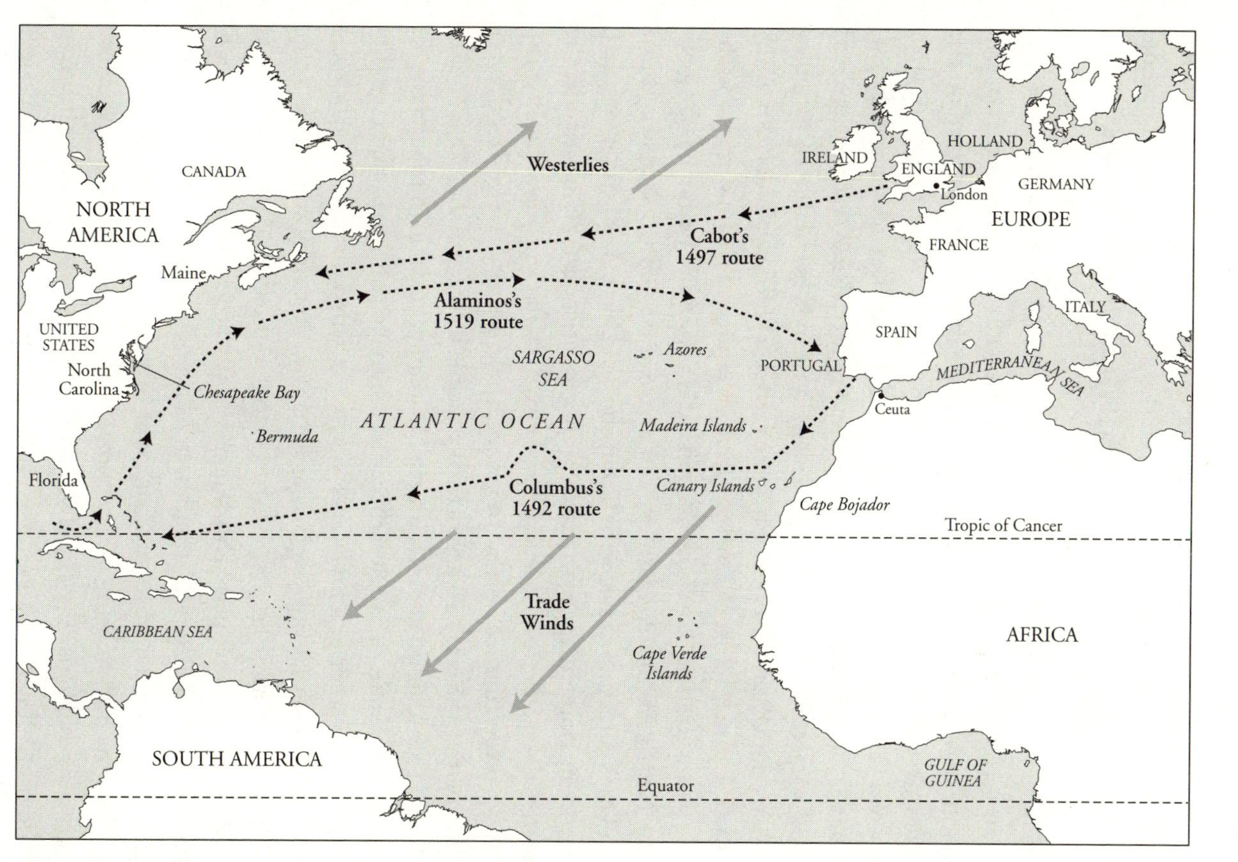

Voyage routes of Columbus, Cabot, and Alaminos

been lost. The historical evidence is scanty: some letters and maps by men who did not sail with him. As a result of this secondhand reporting, there are many conflicting theories about the true nature of his trek across the Atlantic.

What we do know is that Cabot's ship, the *Matthew*, was a *navicula*, meaning a relatively small ship of fifty tons. It had a high rounded stern and was square-rigged on the foremast and mainmast and lateen-rigged on the mizzenmast. With this combination of sails, the *Matthew* could sail with the wind, but when conditions changed, it could beat to the windward. What she lacked in size, she made up for in agility. The *Matthew* departed Bristol sometime in May 1497; scholars have not been able to pinpoint the exact date. But most of them agree Cabot would have sailed down the Bristol Channel, across to Dursey Head, Ireland, and then north along the west coast of Ireland. But how far north did Cabot sail before heading out to the North Atlantic? Again, we can arrive at no definitive answer, but a John Day's correspondence with Columbus, discovered only in 1956, places the turning point at approximately 54° north latitude. At this latitude, the North Atlantic Current, which flows near Ireland, may have pushed the *Matthew* even further north. To Cabot, the current was a godsend; after all, his plan was to stay as far north as possible to reach Asia.

But Cabot wouldn't press his luck; ultimately, he decided to set a course westward. In spite of fair weather and relatively calm seas, the crew became anxious after several weeks at sea. At this latitude, the *Matthew* was only slowly making progress against the prevailing westerlies. Some thirty-five days after leaving Bristol, Cabot sighted land. But where was his landfall? Many experts think it was Nova Scotia, but others strongly argue for Newfoundland, Labrador, or Maine. If we hold to the belief that Cabot maintained a westward course at or near 54° throughout the voyage, then his first sighting of land would have been Labrador. But before making landfall, his tiny vessel would have been carried down the coast by the south-flowing Labrador Current. Along this path, they may have intercepted Cape Breton Island, which seems to be the consensus choice among historians for Cabot's most likely landfall. Regardless of the exact location, Cabot was the first European since the Vikings to set foot in this territory. Aided by strong west winds and the Gulf Stream, Cabot's triumphant return voyage to Europe would take only fifteen days. Back in England, Cabot was promoted to the rank of admiral, and King Henry awarded him another patent for a new voyage. In 1498, Cabot, still intent on finding a westward passage to Asia, again departed from England, with five ships this time. He would never make it across the Atlantic; Cabot and his expedition were lost at sea, never to be heard from again. But European nations' hunger for the riches of the New World was just beginning.

The sea paths to the New World that Columbus had blazed would forever transform the Caribbean, bringing migration, colonialism, and exploitation. The trinity of gold, sugar, and slaves would be the impetus for the establishment of settlements in the West Indies by the Spanish, Dutch, English, French, and, to a lesser extent, other nations. Though Spain was the first to establish a colonial system and trade monopoly in the Caribbean, based upon papal authority, Spain's rivals wanted their share. And no struggle was more intense than that for gold.

The Spanish monarchy, hoping Columbus would discover lands rich in precious metals, had underwritten his maiden voyage. They were soon disappointed with his news. Though Columbus did find natives adorned in gold jewelry and heard natives spin tales about gold-laden islands, he returned to Spain with more stories about gold than gold itself. His narratives must have convinced his employers, however, because they would send many more expeditions to the Caribbean, Central America, and Peru in search of gold.

While Columbus discovered the route to the New World, the Spaniard Juan Ponce de León discovered the way back. Though part of Ponce de León's fame rests on his proposed search for the fountain of youth, he sailed for gold. As with most explorers of this period, he was not averse to enriching himself in the wealth of the New World. On March 3, 1513, he sailed from Puerto Rico in the company of Antón de Alaminos. Ponce de León was an adventurer, not a pilot; he relied on Alaminos to be his navigator. Alaminos had sailed with Columbus on his second voyage, and though little is known about his experiences during that voyage, the youthful Alaminos willingly absorbed the piloting knowledge of Columbus, who he respectfully called "the first admiral."

With the indispensable help of two natives who were familiar with the Bahamas, Alaminos piloted the fleet around the dangerous Bahamian reefs. After sailing northwestward for a month, Ponce de León sighted what he thought was a large island, naming it La Florida, "The Flowered One." During his reconnaissance along the Florida coast, he noted in his log "a current such that, although they had great wind, they could not proceed forward, but backward and it seems that they were proceeding well; at the end it was known that the current was more powerful than the wind." This written account confirms that Ponce de León, aptly aided by Alaminos, had "discovered" the Gulf Stream. On this expedition, Ponce de León's three ships crossed the Gulf Stream no less than four times. One of the ships, a large brigantine, attempted to anchor in the deep current and was soon "carried away by the current and lost from sight although it was a clear day." The discovery of the Gulf Stream would play a major role in the Spanish treasure fleets' success at sailing back to Spain with their riches. Before the Gulf Stream became the major conduit to Europe, Spanish vessels typically sought out the midlatitude westerly winds. But ships departing the Bahamas would often struggle to locate these winds, hoping to avoid the ever-present danger of being becalmed and having their voyage severely prolonged. The Bermuda high, a vast and semipermanent subtropical pressure system, was a particular bane to sailors who were unfortunate enough to sail into its band (at approximately 30° latitude) of weak and fickle winds, known to mariners as the "calms of Cancer."

The stature of Antón de Alaminos as a skillful navigator would rise markedly over the years. In contrast, the Spanish monarchy viewed Ponce de León's discovery of Florida as a modest success — no gold or youth-giving springs were ever found. Even his encounter with the Gulf Stream did not merit much praise; it would fall upon Alaminos to demonstrate its importance in navigation. Ponce de León would make one more voyage to Florida in 1521, but it would cost him his life. With subsequent expeditions along the Yucatan coast and the Gulf of Mexico, Alaminos soon came to the conclusion that a passage north of Cuba now referred to as the Straits of Florida connected the Gulf waters to the Atlantic Ocean.

This information would prove invaluable to the Spanish conquistadors who plundered Mexico for its gold. Upon fitting out an expedition for the conquest of Mexico, Hernando Cortez, who knew of Alaminos by reputation, gave command of the fleet to him. Alaminos's responsibility was to carry dispatches and booty from the military encounters back to

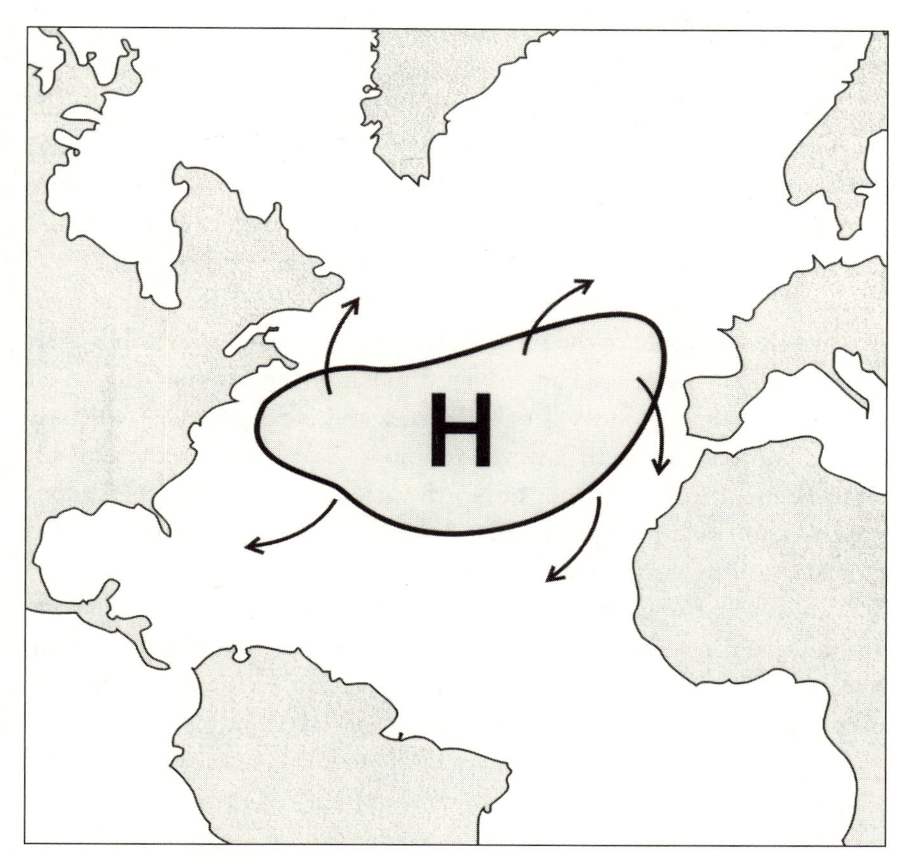

Bermuda high-pressure center

Spain. To Alaminos, this trust in him was a chance to further showcase his navigational skills. He had grown bored with sailing for other captains and longed for his own command.

Outfitted with fastest ship of the fleet, Alaminos sailed from Vera Cruz on July 26, 1519, to Havana. This port city on the northern coast of Cuba was fast becoming the staging area for transatlantic voyages between the Caribbean and Spain because it was ideally located near the Straits of Florida. After leaving Havana, Alaminos's ship was caught quickly in the pull of the Gulf Stream as it flows through the straits. He must have felt that he was in the midst of an old friend, whom he had first visited more than six years ago. The Gulf Stream was to be his constant companion on his voyage along the east coast. But as he sailed northward, Alaminos pondered just how far he could ride this current. His travel with Ponce de León along the Florida coast had terminated midway up the peninsula. What lay ahead? Did the current exit through some unknown egress? Alaminos must have been acutely aware of the *primum mobile* doctrine, which was adhered to by many religious and secular scholars during the sixteenth century. Could he accept the idea that there was no return flow to Europe?

Though the historical record sheds little light on his early life, Alaminos most likely did not have the formal education of the Spanish nobility; he was a self-made man. He had absorbed much of the sea knowledge that Columbus possessed and willingly shared with his protégé. He would rely on this knowledge and his instincts to find his way home. By the time his vessel had reached the latitude of North Carolina, Alaminos's meticulous sightings and detailed plotting of the ship's position must have told him that his heading was no longer due north but rather northeast. The Gulf Stream was carrying him back across the Atlantic. Perhaps on his long journey Alaminos reflected on what he had learned about ocean circulation during his more than twenty years of voyaging. He had sailed the Canary and North Equatorial Currents with Columbus and the Gulf Stream with Ponce de León, and now another current was aiding his eastward progress. Was he able to piece together that he was on the final leg of an interconnected loop of currents, now known as a gyre? It's a tantalizing question, with no definitive answer. What can be said — but has not received wide notice in the annals of maritime history — is that Alaminos was the first European to have sailed the entire gyre. Upon his arrival in Spain, the Spanish were quick to recognize the significance of his feat; he had changed the course of transatlantic navigation forever.

By design, Spanish navigational routes came to follow the gyre pattern in the North Atlantic. The southward-flowing Canary Current had been known for decades; Columbus had sailed upon the North Equatorial Current; and the existence of the Gulf Stream was now widely known and accepted within Spanish circles. In the sixteenth century, Spanish merchants, recognizing the economic importance of the Gulf Stream, commissioned cartographers to produce pilot charts of the current. The results were mixed; mapmakers depicted the current somewhat randomly. In 1525, Diego Ribero produced a map with an intriguing brush stroke that stretched from the waters of northern Florida to North Carolina. Two schools of interpretation have emerged about it: One claims that since early navigators needed pilot charts that marked potential hazards to ships, such as shoals, reefs, or other bottom obstructions, Ribero's paint stroke may have simply been depicting these features. The other speculates that Ribero intended to show the path of the Gulf Stream along the eastern seaboard.

Though other charts that came later would generate still more controversy with regard to their interpretation, one fact became evident: the sea routes to the West Indies and back to Europe were well established. With a lust for wealth, Spain developed a monopolistic economic system in the New World that was heavily dependent upon its treasure fleets, or *flotas*. By Spanish law, the settlers could only trade with authorized Spanish fleets. These ships carried goods for sale to the Caribbean outposts and returned to Spain with precious metals from Mexico and beyond. Beginning in 1561 and lasting until 1748, with few annual exceptions, two fleets, the Nueva España (New Spain) and the Tierra Firme (Mainland), would depart Spain each year to the Caribbean. Following Columbus's route, they would sail down the coast of Africa to the Canary Islands, where the fleets turned westward, riding the favorable trades to the West Indies. More than sixty merchant ships plus armed galleons, which served as protection for the fleet, made up these convoys.

Though the two *flotas* often traveled together, they would separate upon entering the Caribbean: the Nueva España bound for the Mexican port of Vera Cruz, and the Tierra Firme headed for the South American ports of Cartagena, Nombre de Dios, and Porto Bello. In these ports, the ships traded their goods for the wealth of the Indies and South America. When the vessels were loaded with their respective treasures, they rendezvoused in Havana to be refitted and replenished for the long voyage to Spain. But the return trip often proved to be more dangerous than the captains and crews had bargained for. The Gulf Stream was the way home, but the piloting skills of the navigator were put to the test because the current skirts the reef-strewn Florida Keys. Many of these treasure ships, fully loaded to the point of being almost unseaworthy and manned by a crew that was tired and often plagued by illness, wrecked upon those reefs. None of these ships has received more notoriety than the *Atocha*, which sank in just fifty feet of water, with a treasure of forty-seven tons of silver and gold and millions of dollars worth of jewels. This great treasure remained inaccessible for hundreds of years until 1985, when salvager and entrepreneur Mel Fisher recovered it. Though the salvaging of the *Atocha* received considerable attention from the media, modern treasure hunters believe that much of the Spanish wealth looted from the West Indies over a two-hundred-year period still resides on the seafloor underneath the Gulf Stream.

During the seventeenth century, an industry known as "wrecking," or salvaging goods from wrecked ships, emerged in the Keys and Bahamas, and it continued well into the nineteenth century. Wrecking was not a full-time occupation for most of its participants, many of whom were spongers or fishermen who happened to be in the right place when the opportunity arose. Of course, the salvagers expected to be compensated by the owners of the unfortunate vessel. But when the parties could not arrive at a mutual agreement, the legal system determined the appropriate compensation. Records show that in one case of a sunken ship the so-called salvagers removed and consumed a considerable portion of its cargo — beer. The court ruled that no additional compensation was justified.

By the end of the seventeenth century, Spain's restrictive economic system in the New World began to crumble. As a result of the expansion policies of France's Louis XIV, two major wars, between 1688 and 1715, set Europe into turmoil. Spain in particular suffered greatly. Between 1702 and 1708, the English naval forces sank or captured two of Spain's treasure fleets. Unable to keep her trade routes open between the New World and the Old, Spain lost her major source of wealth. By the time the wars ended, with only smoldering hostilities remaining, Spain was essentially bankrupt and in dire need of a financial infusion. In 1715, King Philip V of Spain hastily dispatched a fleet to the West Indies in order to resume Spain's lucrative trade route. The eleven ships making up the fleet arrived in Havana during the summer and immediately set about the task of loading the gold and silver that had been accumulating during the wars.

The commander of the fleet, Captain-General Don Juan Esteban de Ubilla, was reluctant to depart Havana due to the impending hurricane season. The destruction of another treasure fleet in 1711 by a hurricane off the coast of Cuba was fresh in his mind. But under tremendous pressure from the Spanish monarchy, Ubilla made the decision to set sail for Spain, a fateful decision that would have major repercussions on Spain's financial health. On the morning of July 24, 1715, the ships departed the Havana harbor and sailed northward into the Florida Current to begin the first leg of their voyage. After exiting the Straits of Florida, Ubilla positioned the fleet far enough away from the eastern coastline of Florida to take advantage of the Gulf Stream and yet avoid the treacherous shoals that fringed the coast. After five days of uneventful sailing, Ubilla breathed a sigh of relief. The skies were clear, and light breezes gently rocked the ships. Little did Ubilla know that a hurricane was brewing near the Greater Antilles. But on July 29, Ubilla noticed a sea change — long swells were coming from the southeast. These waves were the forerunners of a churning hurricane now racing northward on a collision course with the unsuspecting fleet.

By the morning of July 30, high-level cirrostratus clouds, which radiated outward from the storm's eye, partially obscured the sun, creating a halo. Ubilla was unsure how to interpret this sign in the sky. Being prudent, he ordered his crews to prepare the ships for the possibility of inclement weather. Throughout the day, conditions rapidly deteriorated. The skies darkened, and the winds gusted to well over thirty knots. With the approach of nightfall, even experienced navigators and crew members became alarmed at the violence of the wind-tossed ocean. The waves had shortened and steepened, cresting at twenty feet. Ubilla made one last attempt to keep his fleet from being blown against the looming reefs, ordering all the ships to head into the wind. Throughout the night, the wind speed rose steadily; the ships did not respond to the helm. Ubilla was losing control of his fleet. By the early morning, the hurricane was upon them, driving the entire fleet onto the Florida reefs. In the harsh daylight of July 31, those that had survived surveyed a grim scene: every ship had broken apart on the reefs, and more than a thousand bodies, including Ubilla's, littered the shoreline. News of the disaster would spread quickly to Europe, and though Spain would send a salvaging crew to recover the treasure, much of the gold, silver, and jewels remained untouched for centuries on the ocean floor. Spain was sent reeling from another blow to its economic empire in the New World.

Spain's problems in the seventeenth and eighteenth centuries would take on a more human face, that of piracy in all of its many forms. Piracy, of course, is as old as humanity and inextricably linked with trade and travel. Because of Spain's dependence on the wealth of the New World, coupled with her restrictive policy of forbidding her colonists the right to barter with other nationals, her enemies became more numerous. Almost from the beginning of Spain's presence in the New World, sailors from other nations began to attack Spain's interests, a threat that would remain for generations to come. Initially, they came from Europe itself, sailing the trade routes, which were well known in navigation circles by this time. The French, who were almost always at war with Spain during its reign in the New World, were the first to attack Spain's Caribbean ports. The French became so accustomed to traveling the North Atlantic gyre that by the mid-1500s as many as thirty ships raided the West Indies each year, virtually controlling the seas and inflicting severe economic losses on Spain's colonies. Jean-François de Roberval epitomized the brashness of the French pirates. Though he was born a nobleman, his mounting debts, incurred from leading the expensive life of a courtier, forced him into the lucrative world of piracy. In 1543, he attacked the Spanish stronghold of Santa Marta in Colombia, followed by another foray into the Colombian port of Cartagena in 1544. Not satisfied with these plunders, de Roberval set his sights on the Caribbean. In 1546, his ships attacked Havana — the epicenter of Spanish dominance in the New World — and looted the city. Despite all of his success, de Roberval was not able to regain his wealth; he died a pauper. But the riches of a pirate life would attract a new brand of adventurers.

Early in the seventeenth century, a group of freelance traders, predominately French, who banded together became known as "buccaneers," from the French term *boucanier*. Originally settling illegally in Hispaniola (the island now shared by Haiti and the Dominican Republic), they derived their name from their method of smoke-curing meat on a *boucan*, a combination fire pit and grate. The buccaneers were skilled hunters who killed their own meat, which they traded along with hides to passing ships in return for tobacco, powder, shot, and clothing. Viewing these settlers as dangerous intruders, the Spaniards systematically drove them off the island in 1630, but many of these displaced hunters found refuge on the small island of Tortuga, two miles off the northwest coast of Hispaniola. Unable to maintain their hunting lifestyle on Tortuga, the hunters became sea raiders. Tortuga was strategically located near the Windward Passage, a sea-lane between Cuba and Hispaniola. Operating from small, shallow-draft boats, the buccaneers preyed upon Spanish ships that used the passage to expedite their journey from their Central American ports to the Atlantic. Among Spanish sailors, the buccaneers quickly earned a reputation as fearless and often reckless fighters capable of chilling cruelty who expected no mercy and gave none. To understand the life of the buccaneers, one has to comprehend the times in which they lived. Seventeenth-century societies tolerated far more cruelty than modern-day civilizations: torture was commonplace, abductions were daily occurrences, and death sentences were routinely handed down for minor infractions.

The buccaneers sailed and looted under a strict set of rules, the principal one being "No prey, no pay." After plundering a vessel, they pooled and equally divided the booty, but extra shares might be given to those men who were severely wounded during the encounter. By 1665, the buccaneers had seized so many Spanish ships that their fleet had grown to the size of a national navy. They were so successful in inflicting economic hardship on the hated Spanish that seamen from other countries were enticed to enter into a life of raiding and pillaging. The English buccaneers selected Cagway (Port Royal), Jamaica, as their base of operations. During its heyday between 1665 and 1675, the port became a bustling center for pirating, trading, and smuggling and one of the richest settlements in the Caribbean, attracting all types of rogues and scoundrels.

Henry Morgan, a Welshman who arrived in 1655 in Jamaica to fight the Spanish guerrillas in the island's highlands, became perhaps the most famous of all the buccaneers. He had honed his seafaring and leadership skills during a two-year buccaneering expedition against Spanish interests throughout the Caribbean. Between 1665 and 1671, English buccaneers, under Morgan's "admiralship," raided Maracaibo, Porto Bello, and other outposts on the Spanish Main, culminating with the devastating sack of Panama. Morgan assembled more than two thousand buccaneers in three dozen ships at Cape Tiburón on the southwest coast of Hispaniola for his campaign against Panama. Ignoring rumors that England and Spain had signed a peace treaty, Morgan sailed to the Chagres River on the east coast of Panama. After capturing the fort that protected the entrance to the river, Morgan and his men undertook a grueling, nine-day march to Panama City. Overcoming a vastly larger force, the buccaneers plundered the city, leaving most of its homes and shops charred ruins. Morgan returned triumphantly to Jamaica with booty of more than two hundred slaves and four hundred thousand pieces of eight. Though originally punished by the British for violating the truce between England and Spain, he gained the favor of Charles II, who knighted him in 1674 and sent him back to the Caribbean as a rich man. The buccaneers' rise to power, as epitomized by men like Morgan, during the seventeenth century helped to weaken Spain and, in turn, shaped the history of the West Indies.

Most likely, the buccaneers were the forerunners of the early eighteenth-century pirates who roamed the West Indies, but lineage becomes blurred and titles less clear-cut over time. While buccaneers attacked only the Spanish, pirates were the quintessential sea-robbers, attacking and stealing from all and pledging allegiance to no nation. But during the time of war, declared or undeclared, governments would commission privately owned vessels (privateers) and crews to capture enemy merchant ships and to commit other hostile acts that would otherwise be condemned as piracy. "Letters of marque" licensed sailors to plunder foreign ships without being charged for piracy, which was punishable by death. Henry Morgan always sailed with a commission as a privateer, but to the Spanish, he was nothing more than a pirate.

England has recognized Francis Drake as one her greatest sailors, but he was also a privateer in Her Majesty's service. Around 1568, Drake sailed from Africa to the Spanish Main with a cargo of slaves. While attempting to unload his slaves at San Juan de Ulúa, he encountered severe resistance from the Spanish, who regarded him as an enemy and a trespasser on their territory. The Spanish warships mustered an all-out attack on Drake's ships, which they scuttled. Drake survived the attack but hated the Spanish ever after. During a ten-year period, he made no fewer than seven devastating raids on Spanish outposts. His most celebrated adventure was the capture of the Spanish Silver Train in Panama. With a crew including French privateers and Maroons — African slaves who had escaped their Spanish captors — Drake sailed to the waters of Panama in March 1573. After securing the coastline, Drake tracked the Silver Train to the port city of Nombre de Dios. Catching the train's guards by surprise, he quickly took control of the train and its cargo. He sailed back to England a wealthy man.

Pirates often recruited from the ranks of former privateers. Whenever a war ended, these men found themselves without gainful employment and so were more than willing to pursue the prospects of quick gain, a rogue's life, and escape from the harsh discipline found on naval or merchant vessels. Piracy reached its peak in the West Indies during the early eighteenth century; some historians estimate the number of men willing to sail under the "Jolly Roger" banner as high as five thousand.

In 1651, the English parliament enacted the first of the Navigation Acts, which restricted its Caribbean colonies to trading only with the motherland and to exclude all other nations. This statute turned out to be a major blunder because it led to the swelling of the pirates' ranks. The English had not learned from Spain's earlier mistake of attempting to control all facets of colonial commerce. Pirates were more than willing to sell, at reduced prices, to the colonists the goods and products taken in their raids. So successful were these commercial ventures that the bloodthirsty sea-robbers of the Caribbean also became some of its busiest "traders."

The lure of wealth and adventure attracted numerous nationalities to the ranks of pirates. Dutch pirates were known as *vrijbuiters*, which combines *vrij*, meaning "free," plus *buit*, meaning "booty." *Vrijbuiters* was later corrupted into the English "freebooters," literally, one who freely plunders or pillages, and the French *filbustiers*. Ultimately, the Spanish word *filibustero*, derived from the French word, found its way back into the English language as "filibusters," who, strictly speaking, were not pirates, but adventurers engaged in a military action, such as a coup or revolution, in a foreign country. The word ultimately landed in the United States Congress in the mid-nineteenth century as a tongue-in-cheek label for a congressman who through his interminable rhetoric would obstruct the proceedings and figuratively hold hostage his colleagues.

Regardless of the nomenclature applied to these swashbuckling rovers of the Caribbean, the pirate ship was at the core of their success. The vessel of choice of West Indian pirates was the single-masted sloop, built in Jamaica and Bermuda. Known for its speed and agility, the sloop could easily overtake a merchant ship and escape pursuit by the heavier man-of-war. Shipbuilders rigged the sloop with both square and lateen sails to increase its maneuverability, and if more speed was needed, the crew hoisted a topsail to take advantage of the generally persistent trades. Lurking within the web of islands, cays, and reefs of the Bahamas, these ships relentlessly preyed upon Spanish shipping. Because of its harbor and proximity to the

Straits of Florida, the Bahamian island of New Providence became a nest of pirates. But the myriad of islands offered more than staging areas for attacks on the treasure fleets, proving also to be ideal sites to carry out the laborious process of "careening" (from the Latin word for "keel") the ships — removing unwanted marine growth from the bottom of the vessels. Any type of infestation, such as barnacles or worms, could adversely affect a ship's performance, and pirates were generally meticulous in maintaining their vessels. To begin the process of bottom-scraping, the captain would sail the shallow-draft sloop right up to the shore and beach it during high tide. As the tide receded, the crew hurriedly got to work restoring the vessel's seaworthiness.

The era of piracy in the Caribbean, which began in the 1560s, reached its pinnacle in the 1720s through the exploits of Bartholomew (Black Bart) Roberts. Born in Little Newcastle, England, he became the most successful pirate in history, capturing more than four hundred ships in the span of only three years (1719–22). Historians tend to overlook Roberts, deferring to more flamboyant personalities like Blackbeard (Edward Teach), who ironically captured fewer than thirty ships. Since he was a meticulous man both in his appearance (clean shaven and well-dressed) and demeanor (always treated all on board with respect), Roberts was not typical of the pirates who roamed the Caribbean. In addition to using surprise and speed to attack unsuspecting vessels, he relied heavily on his navigational skills, which he acquired while sailing the slave routes between Africa and the West Indies. He came to know the winds and the currents of the tropical Atlantic, studying them in a manner that was totally foreign to his predecessors. Confident that this knowledge would give him an advantage, Roberts led his crews in brazen attacks on ships that often had him outgunned and outmanned. With Roberts's death in 1722 off Cape Lopez, Gabon, the curtain was closing on the reign of the Caribbean pirates. The nations of Western Europe with colonies in the Americas began to vigorously protect the sea-lanes that linked the Old World to the New.

8 COLONIZATION OF AMERICA

THE SETTLEMENT OF AMERICA during the seventeenth century was a logical extension of the period of exploration and discovery of the preceding century. While religious and political aspirations were certainly factors in this colonization movement by the English, the lure of gold chiefly motivated England to look westward. The acquisition of new lands to rival the rich mines of the Spanish Main and Peru seemed to the English merchants long overdue. And like their European counterparts, the English viewed the sea as a great highway to reach these lands. As an island nation with strong ties to the sea, England was swayed by all this talk about discovery and colonization. Spurred on by the unfathomable riches in gold and silver that Spain was bringing back from its colonies, England did not stand aside for long before embarking for the New World. The land between what is now Florida and New England was relatively untouched by other European nations and, thus, was ripe for colonization.

The English would have to overcome two main obstacles to achieve their wish of widespread colonization of America: the dearth of seaworthy ships capable of making the transatlantic voyage and ignorance of the navigational techniques Prince Henry had developed and refined more than a hundred years earlier. Elizabethan ships were small and contained little usable space for supplies and passengers. These ships had to carry dozens of apprehensive settlers three thousand miles across the sea, and the colonists would need to take clothing, seeds, tools, building materials, arms, and ammunition in addition to food and water. Apart from their training in England's coastal waters, many captains had little or no open-ocean experience. They would have to rely on their instinct, fortitude, and prayers to

Square-rigged ship (copyright Holger W./Shutterstock.com)

ensure a safe passage. Compounding the problem of a transatlantic voyage, Elizabethan ships were commonly square-rigged, so they were at their very worst when they had to go against the wind. Consequently, English captains would often remain in port for days or weeks at a time until favorable weather conditions allowed for a hasty departure. For England, colonies were an expensive and risky investment.

But the strong westerly winds that are common to the latitude of England could not deter Sir Humphrey Gilbert, an English explorer intent on finding a sea route to Asia through North America. Though a visionary, Gilbert lacked practical seamanship skills. On June 11, 1583, Gilbert set forth with five ships from Plymouth, England. With the prevailing westerlies his constant nemesis on the voyage, Gilbert tacked from one latitude to another , jockeying his position to find more ideal conditions. After seven arduous weeks, during which one ship had to return to port because of leaks, Gilbert finally made landfall at Newfoundland. Though he claimed the land for Queen Elizabeth I, the barren landscape and harsh climate deterred England from establishing a permanent settlement until 1621. It may appear puzzling that Gilbert had chosen a northerly route to reach the New World in light of Columbus's success using a southern passage with favorable trade winds and the North Equatorial Current. The English Crown had authorized Gilbert to establish a settlement anywhere in North America, and he was also well aware of the existence of the Gulf Stream. Did his zealous quest for a Northwest Passage cloud his judgment? (In 1566, he had written *A Discourse of a Discoverie for a New Passage to Cataia* [China] in which he urged the queen to seek this route to the Orient.) Or did his practical side win out? Gilbert was cognizant that the Spanish and the Portuguese controlled the southern route and that a northern one might be the best way to avoid any confrontation that could endanger the expedition. His first attempt to reach America had failed when the stormy North Atlantic forced him to seek refuge in Plymouth, England, and his financial backers would not be likely to look favorably upon another failure. We may never know his true motivation, because on the return voyage to England, his vessel, the *Squirrel*, shipwrecked, with the loss of one hundred lives and many of Gilbert's records. He was last seen on the deck of his floundering vessel, a book still in his hand.

Although in the modern world detailed nautical charts are readily available to anyone who wants them, during the sixteenth century, knowledge of preferred ocean routes was not commonly shared among competing maritime nations. A pilot who was fortunate enough to find his way to the West Indies, Panama, or Peru was not prone to share this hard-won information with anyone except possibly an employer. As the English became more determined to colonize America, they needed expert navigators to take them there. But these navigators belonged to an elite and exclusive fraternity of men who had made the transatlantic voyage. England, as a second-rate naval power at best, would have to go it alone.

The year 1584 marked a turning point in England's attempt to colonize America. Sir Walter Raleigh, the half brother of Gilbert, petitioned Queen Elizabeth to establish an American settlement beyond the reach of Spanish influence so that English privateers could harass Spanish vessels loaded with the wealth of the Spanish Main. The mid-Atlantic coast appeared ideal for this venture since Spanish ships sailed northward via the Gulf Stream for their return voyage to Spain.

Raleigh prudently decided to send out only two reconnaissance ships, which were charged with exploring this territory. Under the commands of Captains Philip Amadas and Arthur Barlowe, the ships departed the Thames on April 27, 1584, and immediately set a course directly to the south, as opposed to Gilbert's westward trek. In retrospect, this was a bold move because of the dearth of nautical charts available to them. English pilots had only scraps of navigational information: a general whereabouts of the West Indies and knowledge of Columbus's voyage to these islands by way of the Canary Islands. Their proposed circuitous route to the New World defied the nautical logic of the day: a straight line drawn from England to Virginia would represent the shortest route. Amadas and Barlowe were confident, however, that the Canaries route provided a major advantage: ease of navigation. While sixteenth-century mariners could, in theory, determine latitude with some degree of accuracy, the ability to measure longitude was still more than 150 years in the future. But any sailor worth his salt could determine direction. Like many ships of this era, the vessels of Amadas and Barlowe were equipped with compasses. (The time and place of the compass's origin, as well as when mariners first applied it to navigation, are unknown.) The compass was virtually all that Amadas and Barlowe needed to successfully navigate the Canaries route. "Sail south till your butter melts" was the old mariner's quip that guided them.

The two ships reached the Canary Islands on May 10 and immediately struck a course westward, driven by the gentle push of the northeast trades. Fortunately, the crossing was uneventful since the hurricane season was just beginning. Their luck continued to hold as they entered the Gulf Stream off of Florida on July 2 and sailed northward under its incessant pull. Standing on the decks of their ships, they scanned the western horizon, not knowing what to expect from this strange new land that was now tantalizingly close. As they neared land, they perceived a fragrance coming off the coast "as if they had been in the midst of some delicate garden," teeming with all kinds of fragrant flowers. Though tempted to make landfall, Amadas and Barlowe pressed on to the barrier islands of

North Carolina. On July 13, they dropped anchor at Core Banks, located between Cape Lookout and Cape Hatteras. It was a historic occasion because Amadas and Barlowe had demonstrated the efficiency and potential profitability of the southern route to North America. Just as river travel would open previously inaccessible areas of early America, ocean currents, particularly the Gulf Stream, would now play a major role in the settling of the New World.

After six weeks of additional reconnaissance, during which they traded with the natives, Amadas and Barlowe got underway, riding the northern limbs of the great oceanic and atmospheric gyres back to Europe. Arriving back in England by the middle of September, they had completed a loop around the Atlantic in just less than five months. Amadas and Barlowe gave such a glowing account of the land — the mildness of the climate, the heavily wooded shores, the abundance of game, and the friendliness of the natives — that Raleigh named the territory Virginia, as a testament that the discovery was under the patronage of a virgin queen.

Buoyed by the news from Amadas and Barlowe, Raleigh became more determined to establish a permanent colony along the mid-Atlantic coast. In the spring of 1585, he dispatched a second expedition to North America. Commanded by Sir Richard Grenville, the fleet consisted of seven ships: *Tyger* (the flagship), *Lyon*, *Elizabeth*, *Roebuck*, *Dorothy*, and two smaller pinnaces. Though Raleigh himself owned the two pinnaces, his choice to send them on an open-ocean voyage is surprising in light of their small size. He would soon regret his decision.

Taking the southern route, which would become so widely used by other English vessels in the years to come that mariners affectionately referred to it as the "old course," Grenville's fleet ran into bad weather off Portugal. One of the pinnaces was swamped and sank, and the storm scattered the rest of the fleet. Alone, the *Tyger* sailed on to Puerto Rico, crossing the Atlantic in only fifteen days. (By contrast, Amadas and Barlowe had sailed from the Canaries to the West Indies in thirty-one days.) The hard-won knowledge of when and where to encounter the most favorable trades was beginning to reap dividends.

Upon arriving at Puerto Rico, the crew of the *Tyger* erected a temporary fortification for protection (the island was then a possession of Spain) and proceeded to build a new pinnace. A week later, they were alarmed to see the approach of a ship, fearing it was an attack from the Spanish. But to their relief, the *Elizabeth*, separated weeks earlier from the *Tyger*, had come three thousand miles to rendezvous with them. After taking on supplies, the ships sailed north, the captains secure in the knowledge that information on the Gulf Stream compiled by Amadas and Barlowe was reliable and accurate. They arrived safely off Cape Fear on June 23, 1585, and on June 26 anchored near Wococon (present-day Ocracoke and possibly Ocracoke Inlet) on the Outer Banks. However, three days later, the *Tyger* ran aground; her supplies were lost to the sea; and her crew grew weary of their long voyage. But their spirits were raised by the good news that about thirty Englishmen were on Croatan, near Cape Hatteras. These men had disembarked from the other ships; evidently, all the ships had successfully made the journey.

Without further delay, Grenville and his men, seeking a site for the colony, set out to explore the sounds, marshes, and inlets of this remote area. While the pinnaces had proven vulnerable in the vast Atlantic, they were in their element in the shallow waters of the barrier island lagoons. What they lacked in size, they made up for with speed and maneuverability, particularly in shoal waters. Traveling in these shallow-draft vessels, Grenville's party was able to cover a great deal of territory, including a number of Indian settlements. Moving slowly northward, Grenville finally brought the first colonists to their new home on Roanoke Island. Founded in the summer of 1585, the English colony of Roanoke was established twenty-two years before Jamestown and thirty-seven years before the Pilgrims came ashore in Massachusetts.

Grenville would leave a small band of men under the leadership of Ralph Lane to oversee the building and manning of a fort on the island. His intent was to hastily sail back to England for supplies and return the following Easter. Unfortunately for these men, Grenville could not return on the appointed day, and Lane's men were ill-prepared, and not resourceful enough, to fend for themselves. Though they depended on the local Indians for much of their food, their relations with the Indians began to sour when the colonists grew more hostile and antagonistic toward them. With food running out and Grenville's fleet still in port at Easter, the situation looked grim for the Roanoke colonists. But help arrived in time in the person of Sir Francis Drake, now famous throughout England as an adventurer, privateer, and explorer. Flush with victory over the Spanish in the West Indies and Florida, Drake made a planned stopover at Roanoke Island in June of 1586. (As a superb navigator circumnavigating the globe in 1577, Drake almost certainly employed the Gulf Stream to facilitate his journey northward.) While Drake offered the colonists a month's supply of food and a vessel, the *Francis*, which could carry them all back to England, he would not wait long for Lane's decision; he was anxious to get back to England to claim his prize money from his Spanish raids. But after a storm, possibly a hurricane, forced the *Francis* out to sea and with the threat of attack by the Indians looming over them, Lane had no choice but to abandon the fort and sail with Drake. And so June 18, 1586, would mark the first of a series of failed attempts to colonize Roanoke Island.

Grenville ultimately returned to Roanoke that summer with supplies and more men. But upon finding the fort deserted, he left only a token group of men there. Undermanned and probably lightly armed, this garrison was unable to defend itself from attacks by the Indians, who had now become quite hostile toward the European intruders. While the Indians killed many of Grenville's men outright, they also drove a number of them from the fort. These men were never heard from again.

But there would soon be a second attempt to establish an English colony in the New World. Undeterred by the previous adversity, John White, one of Lane's colonists, petitioned Raleigh to stage another expedition. Though skeptical, Raleigh conceded, and on May 8, 1587, White and 110 other people set sail from England. Arriving at Roanoke Island in August 1587, the would-be settlers were disheartened to see only the ruins of the outpost. No other Englishmen were in sight. Determined to make Roanoke Island a permanent colony, but acknowledging that additional supplies would now be needed, White decided to return to England, leaving the others there for the coming winter. Lane bore his decision with a heavy heart because he had to abandon his daughter, who had given birth to the first child, Virginia Dare, born of English parents in America.

The ongoing feud between England and Spain would delay his return to Roanoke Island, and White became desperate to make up for lost time. As with Amadas, Barlowe, and Grenville before him, the system of currents in the North Atlantic would facilitate White's return. In 1590, he wrote about the portion of his voyage from the Florida Keys to Virginia: "We lost sight of the coast and stood to sea for to gaine the helpe of the current, which runneth much swifter farre off than in sight of the coast." Scholars believe White's account to be the first recorded instance of the approximate position of the axis of the Gulf Stream. (The importance of locating the main axis of the Gulf Stream remains undiminished even today as northbound ships, tugs, and cruise vessels can increase their speed by taking advantage of this strong core flow.) Upon reaching Roanoke in August, he discovered that something had gone terribly wrong — the colony was gone. With his dream shattered, White found the only clue to the colony's disappearance: the word "Croatoan" carved into a tree. Though puzzled, White speculated that it meant the colonists had gone to live with a tribe of friendly Indians in the village of Croatoan on present Hatteras Island. Unfortunately, White was never able to pursue this idea, and though historians have floated a number of theories, including a devastating storm surge that swept through the village, the mystery of the "Lost Colony" is still unsolved today.

The sad saga of doomed expeditions and colonies along the North Carolina coast speaks volumes to us about the unique nature of this area. Though the Gulf Stream comes very close to the chain of North Carolina's barrier islands, early sailors found these shifting ribbons of sand treacherous to approach. In particular, Diamond Shoals, a vast area of submerged sand ridges, was, and still is, a hazardous passage for any ship. Mariners have appropriately coined these offshore waters the "Graveyard of the Atlantic" — the final resting place of more than a thousand ships.

The year 1606 would usher in a new chapter to the story of American colonization. From this date on, the unfulfilled ideas of Humphrey Gilbert and Walter Raleigh would see fruition with the establishment of America's first permanent settlement at Jamestown, Virginia. The individual who stands out most prominently in this endeavor was Captain John Smith. A bit of a rogue, his early career was marked by participation in a number of military actions throughout Europe and the Middle East. Upon returning to England in the winter of 1604–5, Smith became bored and longed for adventure and travel. To satisfy his wanderlust, he became actively involved with plans to colonize Virginia for profit. He soon procured a fleet of three small ships — *Susan Constant*, *Godspeed*, and *Discovery* — for the expedition. They sailed from England on December 20, 1606, and from the beginning, this voyage was marked by unfortunate experiences. Simply, they never should have left port before March because the North Atlantic and adjacent waters are notorious during the winter months for their gales, which the crew and passengers aboard these vessels would soon experience firsthand. When the fleet was only two weeks out of London, the weather took a turn for the worse, forcing the ships to seek shelter in the English Channel for over a month. In hindsight, the best time for crossing the Atlantic via the southern route would have been during the spring and early summer, when ships can intercept the favorable northerly winds near Portugal and ride them all the way to the Canary Islands, where they veer a bit to the northeast. This northeast flow, the trades, becomes locked in below the Tropic of Cancer and continues unabated over a three-thousand-mile stretch of tropical Atlantic.

A square-rigged ship like the *Godspeed* would behave beautifully if the prevailing winds were favorable to push her along. With a brisk wind filling her sails, the spirits of the crew and passengers would soar; no longer would the cold, gray waters of the North Atlantic have a hold on them. Samuel Elliot Morison, writing in 1954, probably best described sailing this part of the tropical Atlantic: "Sailing before the trades in a square-rigger is as near heaven as any seamen expects to be on the ocean. You settle down to a pleasant ritual, undisturbed by shifts of wind and changes of weather. There is the constant play of light and color on the bellying square sails (silver in moonlight, black in starlight, cloth of gold at sunset, white as the clouds themselves at noon), the gorgeous deep blue of the sea, flecked with white caps, the fascination of new stars arise, the silver flash when a school of flying fish springs from the bow wave, the gold and green of leaping dolphins." Familiar with this stretch of the ocean since he had made numerous trips to the Caribbean, Captain Christopher Newport, commander of the fleet, mapped a route that was similar to his predecessors' — sail south to the Canaries, turn right at these islands, continue on a westward course until the islands of the Caribs, then make another right turn, and sail another thousand miles to Virginia.

Upon reaching the Canary Islands, this flotilla did indeed head west, only to encounter the relatively weak and inconsistent trade winds of that time of year. This error in judgment would significantly increase the duration of the voyage. Instead of two months, it stretched out to 144 days.

And a long journey surely exacerbated the harsh living conditions aboard these seventeenth-century ships. In addition to the cramped living quarters (the ships had not been built to carry passengers) and unsanitary conditions (bathing and changing of clothes was unheard of), one of the major ills of an extensive period at sea was scurvy. With symptoms that included extreme lethargy, painfully swollen gums, and severely discolored bruises, scurvy had been the bane of sailors for centuries. The earliest recorded account of scurvy among seamen dates back to 329 B.C., when the crews aboard the fleet of Alexander the Great suffered from it during their voyage from the mouth of the Indus River to the head of the Persian Gulf. Though Dr. James Lind is commonly given credit for recognizing the importance of citric acid in the diets of seamen as a result of an experiment he conducted on an English naval ship in 1747, Balduinus Ronsseus, in a work published as far back as 1564, wrote about a cure for this dreaded disease. He relates the story of Dutch sailors suffering from scurvy on their return voyage from Spain who , by chance, cured themselves by consuming the Spanish oranges that were loaded in the ship's hull. Despite this information, a number of the passengers aboard Captain Newport's ships would suffer from scurvy. Since these tiny ships had limited storage, it is entirely possible that they sailed from England without any citrus fruits, which were not easily procurable in northern Europe at that time, nor obtained any when they made port at the Canaries to take on fresh water. Perhaps Captain Newport did not want to delay setting sail long enough to procure fresh vegetables and fruits. Or the bonus for a fast voyage may have been so attractive that he was willing to risk the health of the passengers. Newport was, most assuredly, not averse to reaping a sizable monetary compensation for his efforts; after all, he had been a privateer, financed by London merchants to raid Spanish ships.

To make matters worse for all on board, Smith proved to be a troublemaker on the voyage, and Newport accused him of concealing an intended mutiny. When the expedition finally reached the West Indies on March 24, the angry crew and colonists were ready to hang him and proceeded as far as constructing the gallows. By the strength of his will and charisma, Smith avoided the noose. Spared for now, but with Newport still planning to execute him in Virginia, Smith and his fellow shipmates explored the islands and took on provisions, including much-needed fresh water. (At the end of long sea voyages, water often turned an unappealing shade of green. Bacteria thrived, contaminated the water, and rendered it undrinkable. As a result, beer often became the beverage of choice.) After making a last stop on the island of Monito on April 9, they sailed northward by way of the Gulf Stream to the North American mainland. After ten days of sailing and not sighting any land, many of the passengers became alarmed; the crew had guessed that it would take only seven days to reach their long-awaited new home. What had gone wrong?

The process of estimating, or dead reckoning, where you will be at a certain time if you hold your speed and course was, and still is, more art than science. Aboard seventeenth-century sailing ships, the navigator reckoned speed by throwing overboard a "chip log" — a wedge of wood about eighteen inches in length, attached to a rope that had equally spaced knots tied into it — at the ship's stern. As the line played out, another crewmember counted the number of knots that passed over the stern while a sandglass emptied for thirty seconds. The number of knots translates into the ship's speed — more knots, more speed. The term "knot" would soon enter into maritime lexicon as a measurement of a ship's speed (one knot equals one nautical mile per hour). If no sandglass was onboard, a crewmember, most likely, would sing a sea chantey that would last for thirty seconds, or so he hoped. If the measurement of speed was fraught with error, the plotting of a course would cause navigators to lose many a night's sleep. Before modern navigational techniques were available, the navigator incorporated his estimation of the effect of wind, current, tide, and helmsman error on the ship's course — no easy task. Even a weak broadside current, unbeknown to the navigator, could push a ship off its intended course. Sailing in the Gulf Stream, the three ships may have strayed off course due to variability in current flow. Since at this time only the scantiest of knowledge of the Gulf Stream resided among English navigators, they were essentially sailing into uncharted waters.

After having been at sea for four months, even a slight delay of a few days was unbearable to many of the passengers. They had had enough of this voyage and demanded that the ships head back to England. Before Newport would concede to their demands, a fierce gale off the North Carolina coast lashed their ships, forcing the crew to furl the ships' sails and terrifying the beleaguered passengers. Their fears mounted upon hearing waves crashing against the nearby shoals, which would surely have claimed the lives of everyone on board. But they survived. Taking their survival as a sign that their fortunes had changed, the weary colonists decided to press on. In the bright sunlight of a spring day, they reasoned that they had come too far to turn back. Clearer heads had prevailed. On the morning of April 26, 1607, the battered ships limped past what is now Cape Henry, and without any further difficulty, navigated the wide passage into the Chesapeake Bay.

Upon landing, Captain Newport opened his sealed orders and was surprised to find that Smith had been designated one of the leaders of the new colony. Newport had no choice but to spare Smith's life. For the weary passengers, committed to the task of establishing a New World settlement, the journey ended on May 13, 1607, with the founding of Jamestown. But a harsh winter, the spread of disease, and attacks by the native Algonquian Indians would test the mettle of the colonists. In the first year alone, 70 of 108 settlers would succumb to the rigors of the New World. The following winter ushered in the "starving time," during which 440 of 500 settlers died in just six months. The original 1607 settlement was hardly a resounding success. But scholars credit Smith with almost single-handedly preserving the first English settlement and convincing the colonists to persevere in their efforts to make a home in their new land. And except for the writings of this energetic and immensely confident man, much of the history of Jamestown would be lost.

After the Jamestown settlement, an emboldened England expanded its reach along the Atlantic seaboard. For any entrepreneur wanting to enter into the business of colonization, it was not possible to simply go to North America and establish a settlement. A patent or license to colonize was necessary, and the initial investment to secure this permission could be quite significant. The Virginia Company, which controlled the English claim in America, called "Virginia" in its entirety, was a pair of chartered companies created by King James I in 1606 for the specific purpose of establishing settlements on the North American coast. The two companies, the London Company and the Plymouth Company, operated with identical charters, but each had control over different territories. King James granted the London Company permission to establish a colony between 34° and 41° north latitude (roughly between Cape Fear and Long Island Sound) and set the boundaries of the Plymouth Company between 38° and 45° (approximately from the Chesapeake Bay to Maine). The overlapping area was open to either company, but their charters forbade the establishment of settlements within one hundred miles of each other.

By the year 1618, a group of English religious separatists, later known as the Pilgrims, began negotiations with the London Company for a patent to settle in the Virginia territory. (The London Company had already given a patent to John Smith.) To the Pilgrims' surprise, the contract specified that they had to settle in a region near the mouth of the Hudson River. Soon after, other distressing news came from London. Robert Cushman, one of their negotiators for the patent, reported that another religious group, sailing from Holland to Virginia during the winter, had experienced the full wrath of the North Atlantic. Upon reaching Virginia, the body count aboard their ship stood at 130 emigrants. Only 50 had survived the crossing. Still determined to make a new beginning, away from the stigma of religious persecution, 102 Pilgrims set sail on the *Mayflower* from Plymouth, England, on September 15, 1620. Departing this late in year was not in their original plan, but due to legal entanglements, they could not secure a vessel during the summer. The long transatlantic crossing would test their resolve to settle in the New World.

The Pilgrims' contract specified that any ship attempting to make landfall north of the London Company's territory must sail along the more direct northern route, as opposed to the now commonly used southern course. The *Mayflower* would have to sail against the prevailing winds and currents. Doubt must have entered into the mind of the *Mayflower*'s commanding officer, Christopher Jones, as to the seaworthiness of his vessel. The *Mayflower* was essentially a cargo ship, which Jones had sailed only back and forth across the English Channel. Typical of her day, the *Mayflower* was a square-rigged vessel, with towering superstructures fore and aft that protected her cargo but made sailing against the wind an even more difficult proposition. Simply, the vessel was not designed to hold a large number of passengers. These tailors, shoemakers, and printers were crammed into tiny, dank, and essentially airless living quarters between the ship's upper and lower decks. Their constant seasickness, brought on by the incessant pitching and rolling of the vessel, would bring ridicule and scorn from the *Mayflower*'s crew. And conditions would only get worse.

On its trek across the North Atlantic, the *Mayflower* experienced a strong gale that threatened the ship and the lives of those on board. William Bradford, a thirty-year-old corduroy worker, provided the only first-hand account of what transpired during this storm, and his report is at best sketchy. But we can probably reconstruct the events, based upon present-day knowledge of North Atlantic storms. According to the Beaufort scale, a violent storm with winds over seventy knots produces exceptionally high waves with overhanging crests; foam blows in dense white streaks, causing the sea to appear white; the rolling of the sea becomes heavy. How would the *Mayflower* fare in these howling winds and storm-tossed seas? In order to save his ship, Master Jones would have had to act quickly. Furling the sails would be his first course of action, to prevent the wind from shredding the ship's sails but also to minimize wind-induced torque on the sails that could twist the ship's masts violently, ultimately snapping them or causing other serious structural damage to the vessel.

As Jones struggled to maintain control of his vessel, a monstrous wave, possibly a rogue wave (an unusually large one), crashed against the *Mayflower* and cracked a beam between decks, causing the ship to leak badly. As water poured into his vessel, Jones seriously considered abandoning the expedition and returning, if possible, to England. Stephen Hopkins, who eleven years earlier had been a passenger aboard the *Sea Venture*, a ship bound for the fledgling Jamestown colony, exacerbated Jones's fears. During a fierce storm, the *Sea Venture* had run aground on a reef near Bermuda — a tragedy that very possibly was the basis for Shakespeare's celebrated romantic comedy *The Tempest*. Hopkins recalled spending four harrowing days at sea, during which the ship "was growne five feet suddenly deeper with water above her ballast." As seawater poured into the vessel, the screams of the women passengers were deafening. Master Jones wanted no part of hearing the cries of his terrified passengers, cowering in the bowels of the *Mayflower*. But to his surprise, the Pilgrims steeled themselves to the task at hand, helping the ship's carpenter repair the leak. With the *Mayflower* now relatively watertight, the ship was able to weather the storm. (In 1957, the *Mayflower II* — a replica of the original ship — found herself in the midst of a North Atlantic storm. Even with all sails furled, the ship's boxy hull kept her shoulder to the wind. She was rock steady in the heavy seas.)

After a sixty-six-day voyage that covered more than three thousand miles, the *Mayflower* sailed in sight of Cape Cod. As a result of having to fight the elements along its entire journey, the ship had traveled at an average speed of only two knots. So thankful was passenger Elizabeth Hopkins for the safe passage that she named her son, who was born at sea, Oceanus.

But Master Jones had to decide whether to continue sailing — they were well north of their intended destination — or find suitable anchorage along the cape's coast. With a favorable northerly wind and a push from the southward-flowing Labrador Current, Jones headed south, confident that the *Mayflower* could easily cover the two hundred miles in a few days. But in the seventeenth century, navigational aids did not exist and charts of the water between Cape Cod and the Hudson River were unreliable. Even for a seasoned captain, coastal sailing in uncharted waters was much more hazardous than blue-water sailing. As the *Mayflower* sailed south, Jones became anxious, commanding a crewmember to make almost constant soundings of the bottom. Unexpectedly and to Jones's great dismay, two events occurred simultaneously: the wind died down, and the water shoaled markedly. The *Mayflower* had sailed into what is now called Pollock Rip — a labyrinth of constantly changing shoals marked by strong tidal currents rushing between Nantucket Sound to the west and the Atlantic Ocean to the east. Modern-day mariners consider it the most dangerous stretch of water north of Cape Hatteras. And with good reason — more than half the wrecks along the entire Atlantic and Gulf coasts have occurred here. If the *Mayflower* remained grounded for long, it would surely succumb to the Pollock Rip. But fate, and probably a good measure of prayers from the Pilgrims, intervened; the wind shifted to the south. Now, able to extricate his vessel from the shoals, Jones prudently decided to abort his attempt to reach the Hudson River. Instead, the would-be settlers spent nearly a month searching for suitable anchorage along the Massachusetts coast, ultimately deciding on a location near Plymouth harbor. As the patent from the London Company was invalid in New England, the Pilgrims drew up a new agreement of governance, the Mayflower Compact, which became the first governing document in the New World. In it, the new colonists legally acknowledged the location of their settlement: "Having undertaken for the Glory of God, and Advancement of the Christian Faith, and the

Honour of our King and country, a voyage to plant the first colony in the northern parts of Virginia . . ."

The plight of the *Sea Venture* and the *Mayflower* would be repeated many times during England's attempts at colonizing the New World. Numerous transatlantic sailings from the ports of England characterized this period of rapid colonization, but many of these ships were simply not sufficiently seaworthy to make the long and hazardous crossing. From the journal of James Fontaine, a passenger aboard the *Dove* bound from England to Virginia, we have an account of how this frail vessel attempted to weather the elements: to lower the ship's center of gravity and lessen the resistance to the howling winds, two seamen had to go aloft to cut away the tattered sails, broken masts, and other rigging. This task had to be performed for fear of the ship floundering in the wind-whipped seas and ultimately capsizing. In spite of the inherent dangers, the heroic efforts of these men prevented the loss of ship and crew alike.

Others were not so fortunate. While historical records do not provide the total number of ships lost or casualties associated with storms, there are some clues. A report prepared by an English parliamentary committee in 1830 bemoaned the losses caused by "the boisterous nature of the weather and the badness of the ships." It continued: "The annual loss of life, occasioned by the wreck or foundering of British vessels at sea may, on the same grounds, be fairly estimated at not less than One Thousand persons in each year." In retrospect, the success of England in establishing a foothold in the New World rested more on the pluckiness of her captains, crews, and settlers than on the seaworthiness of her fleet.

In addition to England, the Netherlands also had a long maritime tradition, albeit confined to the coastal Baltic and North Seas. But it was eager to expand its mercantile reach. The late sixteenth century would mark a turning point in the commercial and maritime interests of the Dutch throughout the eastern and western hemispheres. They had thrown off the shackles of Spanish control in Europe and entered into a period of unprecedented economic expansion. During the seventeenth century, the Netherlands would rise to a world-trade supremacy unmatched by the other European nations. Amsterdam became a world-class financial center, and Dutch ships could be found throughout the seven seas. During this period of spirited Dutch exploration, a series of Dutch mapmakers merged art and science into making maps and atlases that were unsurpassed as navigational tools. In the studios of Hondius, Blaeu, Janson, de Wilt, and others, Dutch cartographers ushered in the golden age of mapmaking. In particular, the Blaeu family produced many fine maps of the Netherlands and the world, culminating in their massive work, the *Atlas Major*. This opulent atlas, published in eleven volumes, contained almost six hundred two-page maps and three thousand pages of text. But none of this august group of mapmakers would stand out more than Gerhardus Mercator (1512–94), whose peers considered him to be the greatest cartographer since Ptolemy. In 1569, Mercator published a revolutionary map projection that would become an indispensable aid to navigation and forever bear his name (Mercator projection). Until Mercator, cartographers could only accurately represent the continents on a globe, but he devised a way to represent the spherical shape of the earth on a flat page. For the first time, navigators could plot compass bearings on these maps in straight lines.

Guided by these maps, the Dutch East India Company established a lucrative trade route with the Spice Islands in the East Indies. Ever mindful of the financial investments of its shareholders, the East India Company engaged the services of Henry Hudson, an English mariner, to find a shorter route to the East. Typical of seamen of this period — lusting for adventure, eager to sail under any flag that promised glory and wealth — Hudson was more than willing to forgo any allegiance to his home country.

In April 1609, Hudson, in command of the *Half Moon*, set a course along the coast of Norway to search for a Northeast Passage. The ship glided along in the northward-flowing Norwegian Current. (The *Half Moon* was one of the speedy, maneuverable vessels known as "fly-boats" by the English.) Upon rounding the North Cape of Norway (above the Arctic Circle), the *Half Moon* had reached the southern limit of the Arctic ice pack. The crew became alarmed that their small vessel would be crushed by the massive ice floes. Threatening mutiny, they demanded Hudson abandon his plans. Sitting in his cabin, Hudson considered his options: press on to the islands of Novaya Zemlya, contending with strong headwinds and even thicker ice, or change course, defying the mandate from his employer. Hudson, like many navigators of this era, had heard rumors of an all-water passage to the west, and Captain John Smith had sent him a letter and a set of maps that depicted a river — possibly a sea — leading directly to the west. Showing Smith's maps to the crew, Hudson convinced them to set a new course toward North America. As with his English predecessor, Sir Humphrey Gilbert, Hudson would also search for the elusive Northwest Passage to the Orient.

As the *Half Moon* skirted back along the Norwegian coast, a series of storms rocked the vessel, blowing it westward for a few days. At the end of May, the *Half Moon* sailed into the Faroe Islands to take on much-needed supplies and water. Robert Juet, a crewman and the unofficial chronicler of the voyage, wrote that in early June the *Half Moon* departed the Faroes and headed southwest to search for the island of Busse, supposedly discovered in 1578, at around 57° north. Though they never found the island, the sailors "were surprised by the force of the current," today known as the North Atlantic Current. At these latitudes, the *Half Moon* struggled to make headway against the current, the slow passage wearing on the weary crew. But the North Atlantic still had more in store for all on board the vessel. Though it was early summer, more storms beset the ship, snapping her foremast and damaging her deck. Finally, on July 2 the *Half Moon* reached the Grand Banks off Newfoundland. Swept southward by the Labrador Current, the ship passed Cape Cod and continued on its journey, sailing as far south as Cape Hatteras. By the end of August, Hudson was growing despondent; his quest for a sea passage to the Orient was proving fruitless. Reversing his course, Hudson set the *Half Moon* on a northward trek, essentially hugging the Delmarva Peninsula. On September 3, Hudson, on coming to the river that would later bear his name, believed that it was not a river at all, but some great arm of the sea, the much-sought-after Northwest Passage. (Smith's maps depicted a large inlet north of the Virginia settlement.) Upon seeing the Palisades, rock cliffs towering five hundred feet that lined the river's western shore, Hudson was elated. He rationalized that these geological battlements, which date back to the Precambrian age, must be the entrance to the fabled passage to the Orient. For the next three weeks, he sailed the *Half Moon* up the Hudson River, as far north as present-day Albany, until the river shoaled considerably. Unable to proceed further upstream, Hudson reluctantly realized his mistake, sailed back downstream, and set a course back to Europe.

During his return voyage, a discouraged Hudson pondered his fate. He had defied his orders from the Dutch East India Company, and his quest for a westward sea route was a bust. What did he have to show for his efforts? A resourceful Hudson would play the only card available to him: extol the virtues of the land, with its lush forests, ample harbors, fertile valleys, and, most important, an abundance of fur-bearing animals. Similar to the tactics of Amadas and Barlowe in their dealings with Raleigh, Hudson would convince the Dutch that the New World had riches of its own.

In 1621, the Dutch founded a new trading firm, the West India Company, specifically targeting the resources of North America. The directors of the West India Company decided to establish a series of outposts and trading centers along the Hudson, Mohawk, Delaware, and Connecticut Rivers. In 1626, New Amsterdam, located at the mouth of the Hudson, became the hub of a profitable fur trade with the Native Americans. Dutch vessels bound for this colony and other settlements adopted the West Indies–Gulf Stream route, after initially sailing south from the Netherlands.

The Dutch were well acquainted with this route. At the end of the 1590s, Spain's interference had prompted the Dutch merchants to search for new sources of high-grade salt outside the Iberian Peninsula. A key product in Dutch trade during the sixteenth century was herring, and to preserve the catch for distant markets, the herring was packed in salt. During 1599, the first Dutch salt-fleet sailed to the Caribbean for salt. The next year an even larger fleet of twenty-five vessels was sent out to the salt lagoons of Punta de Araya, on the coast of Venezuela. From 1599 to 1605, more than seven hundred vessels would sail to the coast of Venezuela and return via a combination of currents in the Caribbean basin and the Atlantic Ocean. With Venezuela as the staging area, these ships would sail northward in the Caribbean Current through the Windward Passage, turn right at Cuba, catch the Loop Current as it squeezes through the Straits of Florida, and ride the Gulf Stream to ports in the Netherlands.

The navigational information that was acquired by the Dutch as a result of the Netherlands-Venezuela link would spur on competition among Dutch seamen to establish the fastest crossing to New Amsterdam. The 147-day voyage of the *Koning David*—which the passenger Kiliaen van Rennselaer would say "lasted too long"—was cut to only 51 days by the vessel *Houttuin*. In spite of these relatively rapid transatlantic crossings and the lucrative fur trade with the Indians, the founding fathers of New Amsterdam had trouble attracting settlers from the relatively prosperous home country. Over a forty-year period, the Dutch would struggle to sustain this colony, ultimately surrendering it in 1664 to the English, who soon renamed it New York.

During the early part of the seventeenth century, settlements of half a dozen European nations dotted the Atlantic seaboard from Acadia (roughly Nova Scotia) to Florida. Along this extended coast, French, Dutch, Swedish, English, and Spanish colonists built forts to protect their interests and established small trading villages. While the English were setting down roots in New England and Virginia, the Dutch still held the Hudson, the Swedes the mouth of the Delaware, the French Acadia, and the Spanish Florida.

From their outposts in the Caribbean, the Spanish made numerous attempts in the sixteenth century to colonize Florida. Almost every mission failed because of a hostile native population. Ultimately, the Spanish gained a foothold and established the first permanent European settlement, St. Augustine, in 1565. Since gold and sugar were not found in Florida, the Spanish turned to the importation and selling of sugarcane in their new settlements.

Gulf Stream boats, known as smacks, commonly sailed the short distance from Cuba to the Florida mainland with a cargo of sugarcane. These small, single-masted vessels had to cross the narrow but potentially dangerous Straits of Florida. Not only did these tiny boats have to contend with the torrent of water flowing through this passage; they were also subject to the fickle nature of Gulf Stream weather. The tropical and subtropical waters are noted for their squally weather, which is often accompanied by heavy rain, powerful electrical storms, and strong winds. A particularly striking phenomenon common to these waters is the "white squall," the culprit of numerous sea stories that has been blamed for many documented tragedies. Lacking the dark, ominous clouds that are normally associated with thunderstorms, white squalls are said to appear out of nowhere, accompanied by a sudden increase in wind speed and rush of white water toward the unsuspecting vessel. *The Pride of Baltimore*, a 137-foot, 121-ton vessel, heeled over and sank in a matter of minutes in 1968 due to a white squall that had reported wind speeds in excess of seventy knots, pushing a twenty-foot wall of water into its starboard side. Recalling Winslow Homer's painting *The Gulf Stream*, we see a scene that was probably a common occurrence on the sugar route to Florida: a demasted smack in the middle of wind-tossed, white-capped seas. While white squalls have assumed an almost mythical status, as in the 1996 movie *White Squall*, scientific evidence points to their strong, gusty winds as a powerful outflow from a deep cloud.

By the end of the sixteenth century, the Spanish and the French had abandoned the area of present-day South Carolina, content to build a series of forts and garrisons in Florida to protect their economic interests. With other nations out of the picture, Charles I of England in 1629 granted his attorney general Robert Heath a charter to all the lands between 31° (the Georgia-Florida border) and 36° (the North Carolina–Virginia border). He unabashedly named it "Carolina," after the Latin form of his own name. Though the territory was never settled under the Heath charter, a number of wealthy sugar planters on the British colony of Barbados became intrigued about the economic possibilities of this land. They would ultimately bring their experience and expertise, gained in a slave-based sugar industry, to the establishment of South Carolina's first cash crop, rice.

In an undertaking known as the "Barbados Adventures," the Barbadian planters sent ship after ship from Barbados in an attempt to establish a colony in Carolina. The voyages from the West Indies to North America proved to be more adventurous than the planters had bargained for. Storms cast ships upon Bahamian reefs; poorly supplied ships ran out of drinking water; people died. Of the hundreds of passengers who sailed from Barbados, only 148 of them, including three African slaves, lived to establish a settlement at Charles Town Landing, located at the mouth of the Ashley River. But Charles Town would prosper as the residents formed alliances and traded with the Chickasaw, Creek, and Cherokee tribes. However, the backbone of South Carolina's fledgling economy was rice. This commodity was valuable for supporting the West Indies sugar plantations, where very few if any food crops were grown. Rice from South Carolina, as well as cod from New England, fed the slaves.

Rice plantations involved significant capital investments and infrastructure. Since rice grows on marshy land, the owners and their slaves had to clear the fields and construct canals, embankments, and floodgates to regulate the flow of water from nearby freshwater sources. As the plantations flourished, the owners invested even more resources into the construction of more elaborate water control systems to take advantage of the flooding and ebbing of tidal rivers, which were needed to inundate the rice fields.

While wars, hostile natives, and simply bad luck impacted the viability of many of the New World settlements, the fury of Mother Nature would also take its toll. For the early European colonists, hurricanes were a new and terrifying experience, unlike anything they were used to in the Old World. Though Europeans were often at the mercy of the unsettled, stormy weather of the North Atlantic, hurricanes posed a new set of problems to them. These storms quickly became woven into the fabric of colonial life, particularly the financial loss incurred from plantings damaged by hurricanes. Unfortunately for the South Carolina rice planters, who invested heavily in this crop, their fields were not spared from hurricanes.

Since hurricane season overlapped with the period of rice harvesting, these storms routinely flooded the fields, laying waste to the crop. Henry Laurens, a wealthy South Carolina planter, reported, "Ye violence of wind and rain damnified some of the Indian corn and rice." He further lamented, "The ripe Crops of Rice have suffer'd very much along the Sea Coast." John Guerard, another merchant, also noted that though some of his rice had survived a hurricane, most of it was worthless. Guerard wrote, "A great deal pounding away to powder, wch is a natural consequence by its being so long weather beaten & lying in the water wch to be sure softened the grain and causes it to moulder away under the force of the Pestle." Even the infrastructure was not spared; hurricanes often destroyed the facilities used in the production of rice. Laurens, again, wrote that a significant portion of his rice plantation "suffer'd by Salt Water breaking over the Banks" during a strong storm.

None of the above comments are surprising in light of what we know today about the impact of hurricanes on low-lying coastal areas. In addition to wind-whipped waves, the accompanying storm surge — an abnormal rise, sometimes several feet, in the ocean level — are the major flooding agents. A storm surge may be particularly devastating when it coincides with normal high tides and rushes across shallow coastal waters. Bays and inlets can funnel and magnify the surge. With its numerous tidal rivers, South Carolina was especially vulnerable to a high surge. As might be expected, the height of the waves and the storm surge are a function of the intensity of the hurricane. As with all strong storms, hurricanes need a copious supply of energy in the form of heat. Large quantities of water vapor, which have evaporated from the warm surface of the sea, condense in towering clouds, liberating the heat that fuels the hurricane. Hurricanes that have come into contact with the warm waters of the Gulf Stream may rapidly intensify. Since the warm water of this current extends to great depths, the surface winds are incapable of stirring up the cold abyssal water that would cause the storm to weaken. With the current's massive heat reservoir intact, hurricanes may rapidly gain strength as they churn landward. Eighteenth-century South Carolinians knew nothing about the intricate relationship of hurricanes with the Gulf Stream; their only concern was surviving the storm and, when possible, recouping their losses.

Storm-ravaged crops were no small matter to the struggling colonies since they depended on the transfer of these goods to Western Europe for their economic survival. But this transatlantic selling of agricultural commodities was only one leg of the intertwined trading that flourished between Europe, Africa, and America during the eighteenth century. The slave trade that developed became West Africa's main export to America, in exchange for which Europe sent its manufactured goods. The labor of African slaves was fundamental to the economic development of the colonies, and as much as 25 percent of the colonists owned slaves. African slave labor cultivated the tobacco farms of Maryland and Virginia, the rice plantations of South Carolina, and the cotton fields of the South.

Though England had established a permanent settlement in America as early as 1607, the English were slow in taking up the slave trade. But by the end of the seventeenth century, England had taken a leading role in the importation of forced labor, numbering some nine thousand slaves per year, which would increase five-fold by the last decade of the eighteenth century. What factors contributed to this phenomenal growth in the slave labor force? England's trade in slaves was firmly based on the theory of mercantilism. This system of political economy, prevalent in Europe after the decline of feudalism, had certain fundamental underpinnings: it depended on the establishment of colonies, the development of shipping and a merchant marine, a favorable balance of trade, and the retention of skilled artisans at home while supplying the colonies with unlimited,

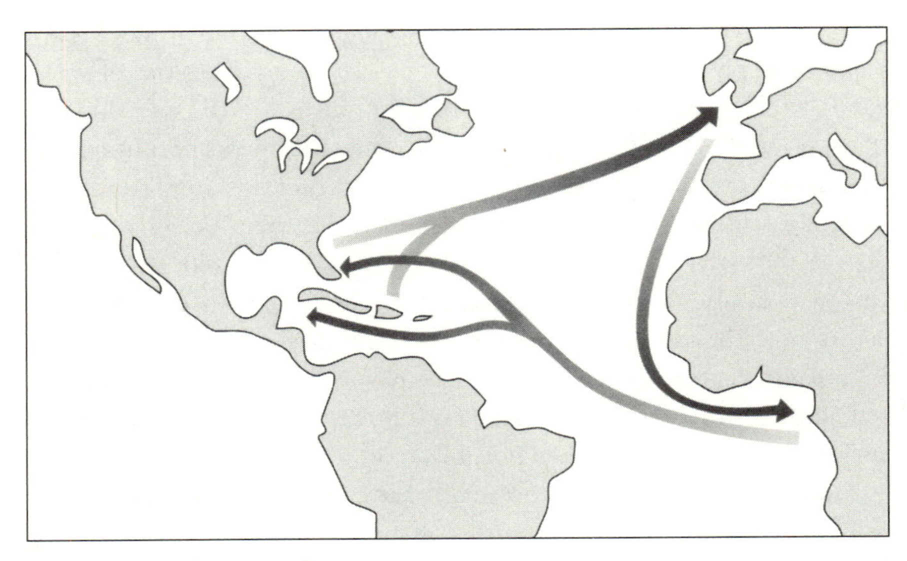

Molasses-rum-slave triangle

unskilled labor. To achieve the latter, an infamous trade triangle linking Europe, Africa, and America would develop across the Atlantic Ocean. Humans would make up the bulk of the cargo.

On the initial leg, the Outward Passage, ships departed European cities, traveled the southern route, and arrived at various ports along the Gulf of Guinea in Africa. While in Africa, merchants traded manufactured goods for slaves and loaded them onto ships to begin the second leg of the triangle, the Middle Passage to the New World. With black Africans stacked like logs, literally one on top of the other, it would prove to be a brutal and dehumanizing experience in which filth, disease, and death were accepted as part of the cost of doing business. Corpses were routinely thrown overboard and the bodies devoured by sharks, which had developed a taste for human flesh. The connection between sharks and human death was so strong that sailors referred to them as "requiem sharks," which came to be used for a number of dangerous species, such as the tiger and bull sharks.

At the height of the slave trade, English shipyards were turning out vessels specifically adapted for the transatlantic crossing. Most of these so-called slavers — the ships that carried enslaved Africans — were built for speed, bulk, and seaworthiness. English slave merchants particularly looked with favor upon ships built in the Liverpool yards, where they were constructing fast and large vessels. In 1797, Wright's shipyard in this coastal city launched the largest vessel ever used in the trade, the *Parr*, a ship of 566 tons that ultimately sailed with a cargo of 700 slaves.

Though slave merchants sought out ships that had the dual characteristics of speed and size, a vessel that could not withstand the rigors of a transoceanic crossing would be of little value. A lost ship meant the loss of forced labor, which was the economic engine of the new colonies. Many slave ships from this period are resting on the bottom off the Florida Keys. The shipboard Africans, who were shackled below, had no chance of surviving a ship's sinking. The Keys' treacherous reefs, lack of reliable navigational information, and the frequency of tropical storms conspired to make this a graveyard for numerous slave vessels.

In addition, the common shipworm was especially troublesome to the wooden slave ships. Christened by sailors "termites of the sea," shipworms are, in fact, not worms at all but clams that are capable of incapacitating a ship. The best-known species is *Teredo navalis*, which thrives in tropical waters and may attain a length of more than two feet. With the ridged and roughened surfaces of its two shells functioning as a tool, this marine mollusk is anatomically specialized for boring into wood. The tenacious shipworm could infest a wooden hull in less than six weeks. Once infested with these organisms, the ship would respond poorly to the helm, slow down considerably in the water, and become more expensive to operate. A badly infested ship could be fatally damaged. With an obvious stake in the viability of these slave ships, Henry Laurens, in an understated fashion, recorded, "The worm has taken possession of the American Brig & it is now condemned." In an attempt to stem the plague of these shipworms, mariners introduced a number of remedies, albeit most of them unsuccessful. As early as 412 B.C., shipbuilders liberally applied a concoction of chemicals, including arsenic, sulphur, and oil, to a ship's hull. By the sixteenth century, the preferred method of protection was to cover the hull with tar and pitch. This treatment was only a temporary "bandage," and the ship often had to be taken out of commission for additional servicing or to replace the timbers. Always cognizant of their financial interest, slave merchants in the eighteenth century were quick to adopt the technique of copper sheathing, which the navies of Europe applied with great success to their vessels. This technology had the desirable effects of reducing maintenance costs and increasing passage speed. As pointed out by the merchant Thomas Williams of Liverpool in 1799, "A still more important saving made by the use of copper on ships carrying slaves from Africa to the West Indies is in the number of lives saved by the shortness of the passage."

Those slaves who survived the arduous transatlantic journey were quickly put to work in the fields and plantations of the New World. Their sweat and toil produced the tobacco, corn, and other commodities that arrived in Europe via the final leg of the triangle, the Return Passage. But during the eighteenth century, the struggling New England colonies would add a new wrinkle to this trade triangle. These colonies would repeat a pattern of activity that had been the savior of many nations during the preceding centuries. They would look to the sea to ensure their survival and economic vitality. With an abundance of suitable harbors, protective headlands, and a reputation for producing seaworthy ships, New England was well positioned to engage in world trade. And New England merchants soon discovered, as had the Spanish, Dutch, and English, that slaves were profitable commodities.

Though Boston merchants were supplying slaves to Virginia by 1678, New England's participation in the slave trade was minuscule throughout the seventeenth century. But by the turn of the century, a number of events would coalesce to change the picture for New England merchants. The influence of the Dutch in the New World would weaken; the English Parliament would severely curtail the monopolistic slave trade of the Royal African Company; the plantation owners in the West Indies needed a new influx of labor. When this combination of events provided the opportunity for New England merchants to be a full partner in the slave business, they did not hesitate. And the trade, as in other instances, would revolve around king sugar, simply the single most important commodity of the period.

By about 1630, most of the Spanish colonies in the Caribbean were no longer sources of precious metals, with Central and South America now assuming a leading role. To survive economically, the colonists turned to sugar. On Columbus's second voyage to the New World in 1493, he brought over sugarcane. Trial plantings occurred on the island of Santo Domingo. Nourished by plentiful sunshine, abundant rainfall, and fertile soil, the crop flourished. Throughout the Caribbean, sugar was to become the "white gold" for European colonists, who would cultivate the crop on almost every parcel of arable land. But with extensive cultivation came the need for labor, and a lot of it, since the planting, growing, and cutting of sugarcane was very labor intensive. Historians have said of the European explorers of the New World that they first fell on their knees to give thanks for a safe passage and then fell on the indigenous population. Columbus fully expected the native Caribs to become willing laborers. Even of his first encounter with these Indians, he reported in his journal, "They are good to be ordered about, to work and sow, and do all that may be necessary." But within eighty years of Columbus's initial contact with the natives, they were a broken people, dispirited by their enslavement and the brutal labor. Their population plummeted, decreasing to an estimated five hundred Indians on Hispaniola from an initial count of between two and three hundred thousand natives. For the sugar industry to survive, enslaved Africans became the new reservoir of labor, constantly replenished by the transatlantic trade.

By 1740, four to five hundred vessels were routinely sailing to the British West Indies and loading their hulls with molasses. As a byproduct of sugar production, molasses is the key ingredient in the distillation of rum, which soon became the economic savior of the New England colonies. A constant convoy of ships would sail northward, via the Gulf Stream, to the rum distilleries throughout New England. In particular, Rhode Island became the hub of rum production, with more than thirty distilleries, twenty-two of them in Newport alone. In spite of their Quaker heritage, Newport merchants would ship millions of gallons of rum to Africa. Ships leaving the protected and deepwater harbor of Newport would trace out a route that was now quite familiar to European mariners: sail eastward in the westerlies and the North Atlantic Current and then southward in the Canary Current to West Africa. Upon reaching the African ports, the merchants exchanged the rum for slaves, now bound to the sugar fields in the West Indies. By the end of the American Revolution, Rhode Island entrepreneurs controlled between 60 and 90 percent of the American trade in African slaves, and Newport was the epicenter in the molasses-rum-slave triangle (with more than nine hundred slaving voyages originating from this port).

While rum may have fortified the colonies, it was that quintessential New England food item — cod — that by the eighteenth century had transformed many communities into thriving centers of international trade.

The proceeds from cod—salt cod, to be exact—financed the growth of towns as small as Provincetown and as large as Boston. Before refrigeration was available, to preserve fish and extend its shelf life, fishmongers salted and dried it. With what appeared to the colonists to be an endless bounty of cod from the fertile waters of the Grand Banks, the curing process was often quick and careless in the rush to keep up with the supply of fish. Upon viewing the salt processing at Provincetown, Henry David Thoreau wrote, "The cod in this fish-house, just out of the pickle lay packed several feet deep, and three or four men stood on them in cowhide boots, pitching them onto the barrows with an instrument which has a single iron point. One young man, who chewed tobacco, spat on the fish repeatedly. Well sir, I thought, when the older man sees you he will speak to you. But presently I saw the older man do the same thing." What was to be done with this cheap and unsanitary cod? Though the best-cured cod was shipped to Spain, the Spaniards would not accept these rejects. With thousands of slaves working the cane fields and all needing to be fed, the solution was to dump the discards on the slaves. The slaves, who subsisted on salted beef from England, would now have to survive on the cod that was disdained by both the American and European communities.

As if the trials of a transatlantic crossing were not enough for the ships that engaged in the slave trade, pirates consistently preyed upon them. Many an unscrupulous mariner found the potential profits hard to resist and signed on to one of the vessels engaged in intercepting and overtaking the slavers. Some became outright slave owners, while others sold cargoes of slaves captured from ships bound for the American colonies. Many outlaws of the sea became a combination of slave merchant, privateer, and pirate; by the 1830s, the term "picaroon" had come to mean both pirate and procurer of slaves. An argument can be made that the growth of the colonies, the importation of slaves, and the ascendancy of piracy were inextricably linked throughout the seventeenth and eighteenth centuries.

Pirates not only plundered the slave ships that they attacked but would routinely commandeer these ships for their own use. Contrary to the myth of pirate ships firing numerous cannon volleys, most of these confrontations involved little bloodshed, were short-lived, and involved little resistance from the lightly armed crew of the slaver. Though slaves and bullion were obvious targets for the pirates, navigation charts detailing preferred sailing routes were literally worth their weight in gold. Aided by these maps, the pirates' sphere of influence soon increased beyond the Caribbean. Pirates cruised the North American coast from the West Indies as far north as Newfoundland, their movements timed to the changing seasons. As Captain Charles Johnson, in his seminal 1724 publication *A General History of the Pyrates*, related, "Pyrates generally shift their Rovings, according to the Seasons of the Year; in the summer they cruise mostly along the Coast of the Continent of America, but the Winters there, being a little too cold for them, they follow the sun and go toward the Islands." Could it also be that pirates were even more weather savvy than Johnson thought? One can speculate that the common occurrence of hurricanes in the Caribbean during the summer and cyclones in the winter, which spawn off the North Carolina coast, controlled or at least contributed to the pirates' seasonal migrations.

Pirates traveled extensively along the North American coast and established outposts in the Carolinas and New England. Charles Vane, who was ultimately hanged for his piratical escapades, operated off the South Carolina coast, much to the chagrin of the Charleston community. Roaming the coastal sea-lanes, Vane looted ship after ship, including a large West African brigantine loaded with ninety slaves and bound for the South Carolina plantations. Though pirates were primarily in the business of trading or selling captured slaves, many pirate captains integrated former slaves into the pirate life, where they often formed a considerable portion of the crew (more than 40 percent of Blackbeard's crew were black Africans). This outward display of racial tolerance and equality would fly in the face of the racial prejudice that New England merchants promoted to justify their slave trade. In spite of all their brutal tactics, pirates of the colonial period often displayed a sense of justice.

Probably no area along the Atlantic seaboard became a more infamous haven for pirates than the North Carolina coast. With its network of protective sounds, shallow inlets, and isolated backwaters, the Outer Banks was a perfect place for piracy to flourish. Of the pirates who used these barrier islands as staging areas for their raids on merchant and slave ships traveling the Gulf Stream, the most notorious was Blackbeard. The little that is known about him, such as his given name (Edward Teach, or Thatch, or Thach), his impressive bearing, his fearless nature, and of course his thick black beard that covered his face, all contributed to striking fear in his victims. But historical evidence suggests that the legend of Blackbeard is more myth than fact since he was a master at perpetuating his image (sticking lighted matches under his hat to give the impression of a possessed demon). The first recorded account of his piratical career is from November 1717, with the seizing of the French slave ship *Concorde* in the eastern Caribbean. After taking command of the vessel, Blackbeard renamed it *Queen Anne's Revenge* (some evidence points to the possibility of his having served as a privateer in Queen Anne's War), which was then one of the largest pirate ships of its era. In consort with other vessels and other pirates, the *Queen Anne's Revenge* would embark on a six-month-long harassment of trade along the eastern seaboard, culminating with the blockade of Charleston in May 1718. Sailing up the coast, the fleet sought the protection of the remote sounds of North Carolina to careen their vessels. Though respected by his crew as a competent sailor, Blackbeard made the fateful mistake of attempting to sail his deep-draft flagship through Topsail Inlet, now Beaufort Inlet. The shifting sands of this shallow inlet made navigation at best a risky undertaking. The inevitable result was the *Queen Anne's Revenge* ran aground, and the ship was ultimately lost in spite of the salvage attempts by Blackbeard and his crew. But Blackbeard would learn from his mistake. After receiving a pardon from Governor Eden of North Carolina for his previous transgressions, the wily Blackbeard used Ocracoke Island as a base of operations to continue his looting and pillaging. Employing Ocracoke Inlet as a conduit to the Gulf Stream's shipping lanes, Blackbeard could virtually monopolize all trade along the mid-Atlantic coast and yet not suffer any military reprisals. The large men-of-war that comprised the bulk of the naval fleet could not safely navigate the shallow and treacherous Ocracoke Inlet.

From its havens along the eastern seaboard, piracy spread its tentacles eastward across the Atlantic and down the West African coast to Guinea, the main center of the thriving slave trade. Pirates would strike at the heart of this lucrative business by using the same system of currents and winds that had proved so valuable to the early explorers, Spanish galleons, and slave merchants. If nothing else, pirates were highly adaptable to the situation at hand. A long ocean passage required the use of a vessel that could remain seaworthy for an extended period of time. The large, three-masted vessel with a full set of square-rigged sails favored by pirates proved to be an excellent flagship for these far-reaching raiders. What it lacked in agility was more than compensated for by its seaworthiness, better gun platforms, and a cargo space more than twice as large as a sloop's. With the use of Caribbean sloops, Atlantic brigantines, and transoceanic square-riggers, pirates roamed near and far, and no part of the vast Atlantic was safe from their predation.

In the short span of a few years, the economic hardships that pirates inflicted on the colonies had become intolerable. There were limits to the insults and depredations that the colonists were willing to endure. In particular, the residents of North Carolina had grown intolerant of the raiding and plundering of their ports and towns. These embarrassing events were so infuriating that the colonists were determined to deal with the pirates once and for all.

However, recognizing the impotence of their own government in dealing with piracy, a group of influential citizens of North Carolina petitioned Alexander Spotswood, the governor of Virginia, to rid the colony of the Blackbeard plague. A staunch enemy of piracy, Spotswood adopted Blackbeard's strategy of using mobile, shallow-draft sloops to navigate Ocracoke Inlet. Commanded by Lieutenant Robert Maynard, the crew of these ships engaged Blackbeard's men on November 22, 1718. During one of the bloodiest battles in North Carolina's maritime history, with neither side granting "quarter" (mercy), Blackbeard was mortally wounded. In what appears to have been a callous decision, Maynard ordered the decapitation of Blackbeard. Though taken aback, the crew carried out his order and affixed the head to the bowsprit of Maynard's ship. Sailing northward on the Gulf Stream to Virginia with this gruesome display essentially signaled the end of pirate supremacy on the Carolina coast.

If the slave trade was at the heart of the New England economic juggernaut during the eighteenth century, then whaling was its backbone. Brave and industrious men, the Yankee whalemen, would become the epitome of all those who "go down to the sea in ships." By 1740, whaling was a flourishing industry; and many products were made from the carcasses of whales: oil for machinery and for lighting from whale blubber, skirt hoops from baleen or whalebone, and whaler's art (scrimshaw) from teeth and bones. In addition, whaling supported the ancillary industries of shipbuilding and sailmaking. At its peak, the colonial whaling fleet numbered 360 vessels, hailing from more than fifteen ports throughout New England and New York.

Whaling lore is rich in tradition, and Herman Melville's *Moby-Dick* is one of the most notable examples; however, New England whalers were merchants above all, not adventurers in the mold of Captain Ahab. They ventured no further in pursuit of these leviathans than it took to meet their quota and fill their ships. In the early days of American whaling, voyages were essentially limited to the whalers' coastal waters and no further east than the Gulf Stream. The relatively wide continental shelf off New England limits the westward penetration of Gulf Stream waters. At these latitudes, the Gulf Stream is approximately eighty miles from the coast of Cape Cod and some two hundred miles out to sea as it flows past Maine. Since extended sea voyages to the whaling grounds were not initially economical, many of the whaling ports developed in and around Cape Cod, close to these grounds. Places like Martha's Vineyard, Barnstable, Dartmouth (later New Bedford), and Nantucket became major whaling centers. Nantucket, which made its fame and fortune from whaling, is little more than a crescent-shaped sandbar rising from the continental shelf, no more than thirteen miles long and about four miles across at its widest point. But by 1715, the inhabitants of this small island were routinely making open-ocean voyages in search of whales.

The original target of these whalers was the North Atlantic right whale (*Balaena glacialis*), so-named because when a ship's lookout spotted this kind of whale, he was to shout out to the rest of the crew that it was the right one to pursue. What made it the "right" whale to hunt was its blubber, which is so rich in oil that it literally floats when dead. Sailing in tiny, single-masted sloops, Nantucket whalers soon discovered that towing a sixty-ton harpooned whale back to port was impractical, if not impossible, considering the unsettled nature of the North Atlantic. Instead, the crew brought the dead whale alongside the ship and stripped away its blubber, which they then sealed in caskets and carried back to the island for rendering.

A member of the cetacean suborder *mysticeti*, the right whale has, instead of teeth, baleen plates made of horny, elastic keratin (like fingernails) that hang down from its upper jaw, which it uses to strain planktonic organisms, mainly copepods, from the water column. During the spring and summer, right whales migrate from their calving grounds off the Georgia and Florida coasts to the plankton-rich northern waters. The longer days and abundant nutrients along the western edge of the Gulf Stream are conducive to major blooms of phytoplankton, which, in turn, support the copepod population. As an ocean-river with no banks, the Gulf Stream's relative position varies from year to year, and, at present, scientists have not figured out how to predict exactly where and when phytoplankton blooms will occur. But some progress is being made using the data from satellites, moored buoys, and research cruises. Scientists speculate that shifts in the whales' location along the New England coast correspond with the geographic variation of prey. Thus, the principle is simple in theory, but sometimes hard to put into practice: find the food, and you'll find the whales.

As American commercial whaling grew dramatically in economic importance during the eighteenth century, merchants and whalers soon recognized the sperm whale (*Physeter macrocephalus*) as the most important of all the species hunted, which included the right, bowhead, humpback, and gray whales. It had the best quality oil, which not only burned cleanly and brightly but was a superior lubricant. Candle makers employed the whale's spermaceti (a white, waxlike substance found in the whale's head) in the manufacture of the finest grade of candles, which became a profitable export to England. In the view of one historian, London was the "best-lit city in the world"; its numerous street lamps flickered brightly with candles produced from the carcasses of North Atlantic sperm whales. On occasion, the intestinal track of one of these whales yielded ambergris, which was extremely valuable as a perfume component, fetching upward of $10,000 for two ounces in the 1880s. As a mate aboard the New Bedford–based vessel *Sunbeam*, bound for a sperm-whaling voyage, aptly put it, "There are only two kinds of whale. One of 'em is the Sperm Whale; the rest of 'em is the other."

Although historical records are a little vague as to the exact beginning of the systematic hunting of sperm whales, whaling chroniclers generally agree that the men from Nantucket ushered in a period of unprecedented growth in the commercial whaling for this species. Legend has it that an intrepid Nantucket native, Christopher Hussey, accidentally came upon a sperm whale in 1712. Though his intended prey were right whales, he killed the whale and brought its oil-filled body back to port — no small task for a single-handed whaler.

By the middle of the eighteenth century, the sperm whale would make New Bedford the whaling capital of the world, replacing Nantucket, which would be disadvantaged by its distance from mainland markets and the small size of its labor force. At the peak of its influence, New Bedford had more than three hundred vessels, valued at more than $12 million, which employed more than ten thousand men — many of them immigrants from the islands of Cape Verde, the Azores, and the West Indies. Whaling crews were always in demand by Yankee captains, who could be ruthless in the treatment of their crewmen and often made their lives unbearable. These captains particularly liked the Portuguese, who had a reputation for being hardworking, cheap, and, most important, eager to sign on to any ship. Azorean males, desperate to escape military service, became the human fodder for many a whaling enterprise. By 1780, more than two hundred whaling ships from New England could be found berthed in the Azores, waiting to fill out their crew. Even escaped slaves found opportunities aboard whaling vessels, which in some cases were manned by all-African crews. Whales and whale hunting would be the underlying factors that made the ports and towns of New England a cultural melting pot. Richard Ellis, noted maritime author and artist, writes in *Men and Whales* that, since the ninth century, men have played an important role in the natural history of whales, as whales have done in the natural history of many lands. Since whales have had a profound effect on humanity, the location of whales generally determined the location of those who would hunt them. Whales, people, and settlements would all be interwoven throughout the colonial period.

Though New England colonists occasionally sighted sperm whales near the coast, they are mainly creatures of the deep ocean, where they routinely sound to depths of more than four thousand feet in search of their favorite prey, the giant squid. The Nantucket whaler Timothy Folger, who was instrumental in helping Benjamin Franklin construct an early chart of the Gulf Stream, was well aware of the association of sperm whales with the deep, fertile western edge of the Gulf Stream. Knowledgeable whalers rarely ventured far from this edge; simply, whales were not to be found in great numbers elsewhere. By necessity, the whalers gained knowledge of the course, breadth, and extent of the Gulf Stream as they extended their search for whales as far south as the Bahamas. They also became painfully aware of the Gulf Stream's strength. Upon the call "There she blows," the crew lowered agile, speedy whaleboats from the whaler (whale ship). With a determined whaleboat crew, these boats were capable of giving chase to a fast-swimming whale. Focused on their intended target, crews sometimes found themselves caught in the swift flow of the Gulf Steam and were surprised to see how quickly they were separated from the whaler, losing all visual contact. Crossing the Gulf Stream involved the utmost care and coordination between whale ship and whaleboat and, at times, could be particularly dangerous. An enraged, harpooned sperm whale could drag its hunters on a "Nantucket sleigh ride," approaching speeds of twenty knots and lasting for two to three hours. It would be a long row back for the exhausted crew.

During the golden age of whaling, the most detailed charts of offshore waters were in the hands of whalers; their livelihood depended on pinpointing the location of the whales. Few, if any, of these charts resided with wealthy ship owners, whom whalers tended to view as untrustworthy to protect their hard-won knowledge. A ship owner would be likely to sell the charts to the highest bidder, regardless of his occupation. As was the case with the early pilots to the New World, whalers could be quite secretive about these charts. Some whalers adorned their charts with caricatures of whales and a written initial indicating the best time to hunt the whale: w (winter), s (summer), a (autumn). Successful hunting simply became a matter of being in the right place at the right time: south in the Atlantic for sperm whales in the spring and then north along the Gulf Stream during the summer, following the right whales.

With these changes in latitude and seasons came changes in weather that could make life miserable, if not downright dangerous, for the whalers. From the blistering heat of the tropics to hurricanes and gales that could transform greasy, slick seas into a roiling mass of water, whalers experienced all that nature had to offer.

Though it might be ephemeral, fog was much more than a passing nuisance to whalers. It markedly reduced visibility on the water, making it difficult to navigate safely between the whaling grounds and homeport. Without the aid of modern navigational devices to plot courses and to detect potential hazards, running aground on shallow shoals and colliding with other vessels were distinct and real dangers. So prevalent was fog a part of New England whaling that Clifford Ashley, an ardent student of whaling and himself a whaler out of New Bedford, depicted it in four of his acclaimed paintings of the whaling era: *Gathering Fog*, *Gray Fog*, *The Sunbeam in a Fog*, and *The Sunbeam Wearing Ship in a Fog*.

Ocean fog forms when very moist air, which has a high water-vapor content, cools to saturation and condenses into liquid water in the form of very small water droplets. Normally, saturation occurs when the relative humidity is 100 percent, but over the ocean, fog can develop with humidity as low as 97 percent due to the salt particles in the air enhancing the condensation process. These droplets are so small and light that they can remain suspended for a relatively long time in the air. The aggregation of these droplets makes up a fog, which is a cloud with its base at the water's surface. In northern New England, cold water, mainly supplied by the Labrador Current, initiates the cooling of the air. During the summer, when warm, moist air from the Gulf Stream flows northward over this colder water, fog is a common occurrence. (On average, New England experiences more than eighty days of fog per year.) Thus, an eighteenth-century whale ship, chasing the summer-migrating whales, would often find itself enveloped by a thick fog, which to the surprise of its crew appeared to have literally rolled upon them.

Though weather and ocean events have conspired throughout history to make many maritime activities, such as exploration and colonization, risky undertakings, the demise of the American whaling industry at the end of the nineteenth century resulted from human activities. The industry was running out of whales because of overhunting. As whalers accumulated knowledge of the whales' habits — such as their reliance on specific ocean current features and preference for specific feeding and breeding grounds — it became easier to find and kill them. So successful was the whaling enterprise during its zenith that of the estimated hundreds of thousands of right whales that once roamed the oceans, only three hundred or so are now thought to exist. Moreover, in a short time, the products of whaling would be replaced by the petroleum industry.

From the sixteenth through the eighteenth centuries, the skill and courage of all those who sailed the ocean shaped the survival and growth of the American colonies. Explorers, captains, pirates, merchants, whalers, and rumrunners would all become intimately familiar with the winds and currents of the North Atlantic. Fleets of brigs, brigantines, sloops, slavers, and men-of-war, to name a few, sailed far and wide in search of new lands and wealth. For the sailors and mariners who crewed on these ships, the era of sailing was rapidly drawing to a close. The golden age of sail in America lasted from about the War of 1812 to the Civil War, when the number and efficiency of commercial sailing vessels were at their peak, in the years immediately before steamships started to take trade away from sail.

At the beginning of the nineteenth century, Maryland became the undisputed leader in American shipbuilding, and Baltimore was the hub of this activity on the Chesapeake Bay. From its shipyards, Baltimore Clippers would emerge as the standard-bearer of a fleet of ships known for their speed, agility, and maneuverability. In an era when speed was synonymous with survival, these ships filled the need for vessels that could elude the powerful but lumbering British naval vessels that harassed American shipping. Baltimore Clippers were "sharp built" — designed with a V-shaped hull that could cut through the water, offering little resistance; as a result, they were fast. But this speed would come with a price; the vessels had relatively little cargo space, a major factor in their ultimate decline. To increase their maneuverability, Baltimore Clippers were gaff-rigged — a configuration of sails that included a triangular fore-and-aft topsail, supported at its peak by a spar called a gaff — that allowed them to sail closer to the wind. In shifting winds, they could easily outmaneuver the clumsy, square-rigged British vessels of this time. Manned by American privateers during the War of 1812, Baltimore Clippers, sailing out of Chesapeake Bay, captured or sank more than seventeen hundred British vessels. Other Baltimore Clippers served as cargo vessels, supplying much-needed munitions and armaments and essentially breaking the British blockade of the American coastline.

Though Baltimore Clippers were instrumental in the survival of the colonies during America's Second War of Independence, their usefulness was short-lived. With the cessation of hostilities, American merchants had little need for well-armed vessels with limited cargo space. The burgeoning colonies needed the reestablishment of transoceanic trade routes and vessels that could carry more goods in a timely fashion. The 1840s marked the introduction of a new generation of fast, large ships that came to be known as Yankee Clippers. Their three masts were fully rigged with square sails, which allowed for very fast downwind sailing. These vessels were capable of reaching speeds of sixteen to eighteen knots, with claims of speeds exceeding twenty knots. The *James Baines* set a transatlantic record of twelve days, six hours, from Boston to Liverpool. Ultimately, hundreds of Yankee Clippers, taking advantage of prevailing winds, currents, and gyres, would roam the earth's oceans. The former colonies would prosper and become the model of economic success for the rest of the world.

During the next century, the maritime nations would switch from sail to steam, but the Gulf Stream would continue to exert its "pull" on the ships that crossed its waters and the souls of the men who sailed these vessels. Over time, the Gulf Stream has yielded many of its secrets, but the deeper that human beings have probed this river in the Atlantic, the more they have come to realize that the Gulf Stream is in many ways still as wild and mysterious as when the ancient mariners first pondered the nature of this current.

EPILOGUE

A COOL, DRY SPRING has made the transition into summer, and I'm on my annual fishing trip to the Gulf Stream. As the charter boat clears North Carolina's Hatteras Inlet and heads east to the fishing grounds, I settle down to the pleasant ritual of watching for flying fish, a sure sign that the Gulf Stream is near. My contemplative mood is interrupted when my son, Lee, blurts out, "When are we going to get there?" The age-old question associated with almost any trip — be it land, sea, or air — gets me thinking about time: the long and the short of it.

Four hundred years ago, a hardy band from England, under the command of Captain Christopher Newport, rode the great atmospheric and oceanic flows to America, setting foot in Virginia. The voyagers founded Jamestown, the first permanent English settlement in the New World. When I attended one of the many celebrations marking the four hundredth anniversary of this historic event, I was fascinated to see a replica of the *Godspeed*, one of the three ships that landed in Jamestown. I became immediately aware of its relatively small size — the cramped quarters must have tested the resolve of even the hardiest seamen on their long voyage across the Atlantic. On my second look at the vessel, I was struck by an even more surprising realization: the *Godspeed* was ideally rigged for sailing the southern limb of the North Atlantic gyre.

Under the incessant thrust of its diesel engines, our boat will soon be motoring into the same Gulf Stream waters that the *Godspeed* sailed centuries ago. Well, not exactly the same waters, I reflect. The Gulf Stream is constantly remaking itself through its meanders, loops, and rings; its glass-flat surface can almost overnight swell with rolling mounds of water.

While the Jamestown settlers may have blazed a trail to America, no visible traces of their journey remain on the Gulf Stream, no sign posts to mark their accomplishments. And yet the Gulf Stream was as important to the settlement and prosperity of America as were the Oregon, California, and Chisholm Trails, which facilitated the westward expansion of the United States populace during the nineteenth century.

Centuries before the first covered wagon left its imprint in the sod of the Great Plains, an eclectic mixture of explorers, slavers, rumrunners, and scoundrels sailed the Gulf Stream in search of new lands and wealth. From Ponce de León, who "discovered" it in 1513, to the Spanish captains commanding the treasure fleets of the sixteenth century, to Henry Morgan, a seventeenth-century English buccaneer who routinely plundered Spain's holdings, the Gulf Stream has floated them all.

A large vessel now appears on the horizon. I immediately recognize it as a container ship — a cargo ship that carries its entire load in truck-size containers. This behemoth, measuring hundreds of feet in length, is steaming northward in the Gulf Stream. It is not unusual to see all types of ships, from small trawlers to large work boats, riding the great saltwater river north. Closing my eyes, I envision the parade of vessels — sloops, schooners, brigantines, caravels, slavers, clippers — that plied the Gulf Stream waters when sailing ships ruled the seas.

The captain throttles the engines back, and I instinctively look at my watch — seventy-five minutes to reach the Gulf Stream; not bad. As we settle into the routine of trolling, I think about the relationship of this ceaseless flow of water to our planet. The Gulf Stream affects and is affected by winds and weather, but also by everything living that it encounters — visitors, sailors, and anglers, as well as flora and wildlife. This river in the Atlantic has a tenacious hold on the human spirit and mind, and it has been the subject of many a myth, story, or tale. Powerful and self-contained, touched on all sides by the unpredictable Atlantic, the Gulf Stream is emblematic of the last vestiges of wilderness on this planet. Henry David Thoreau argued that wilderness allows humans to attain their closest contact with higher truths and beliefs. If, as a society, we continue to value wilderness as a retreat and respite from the pressures of the twenty-first-century world, then how do we protect this watery tract of wilderness? The answer to this question can be elusive because, at its core, a wilderness is a region of bewildering vastness. In this regard, the Gulf Stream is a wilderness of a surprising degree, stretching from the torrid tropics to the polar seas, with boundaries that are fluid, not rigid in space and time. Nonetheless, this is an important question that society must face because we have reached a critical juncture in the long history of the Gulf Stream: the balancing of the exploitation of its resources, particularly its biological resources, due to social progress and economic growth, with the survival of the natural system and its complex web of life.

The Gulf Stream has been around for more than thirty million years. During all that time, it has been, if nothing else, resilient — at times sputtering along, waxing and waning in strength, but always rebounding. Time has taken its toll in many ways: the pirates' cannons have long ago gone silent; the galleons no longer sail the Spanish Main; the whalers are victims of technological change. The Gulf Stream is the only constant in this five-hundred-year relationship between humankind and its mother ocean. This old river keeps rolling on.

Today there is a growing concern that global warming may trigger a sudden climatic shift; that a tipping point will be reached, and the dominoes begin to fall: a shutdown of the Gulf Stream and the whole ocean conveyor belt, latitudinal heat transfer disruption, and the onset of a new ice age. While the topic of global change is at present a contentious issue in both the scientific and political arenas, Jules Verne was, once again, ahead of his time, when, in *Twenty Thousand Leagues under the Sea*, he wrote, "We then went with the current of the sea's greatest river, which has its own banks, fish, and temperature. I mean the Gulf Stream . . . we must pray that this steadiness continues because . . . if its speed and direction were to change, the climates of Europe would undergo disturbance whose consequences are incalculable."

The mate's cry — "hooked up!" — breaks my musing; a dolphinfish has streaked out from beneath a clump of sargassum weed to inhale a ballyhoo. In a matter of minutes, the fish tires, yielding to the mate's gaff; its brilliant colors quickly fade like a setting sun. Out of its element, the fish succumbs to its inevitable fate. I don't feel remorse; all life is biologically wired to live out a certain span — some longer, some shorter. I don't know how many more such encounters I will have with the Gulf Stream, but when I find myself played at the end of the line, I plan to make some jumps and leaps, too.

The writer William MacLeish referred to the Gulf Stream as "the blue god." Is the spiritual undertone appropriate? Possibly. Some view the wilderness as a symbol of innocence and purity. In this romantic model, the wilderness, the last remaining area of unspoiled and uncorrupted nature, is identified with Paradise and the Garden of Eden. I like this interpretation; I'm inclined to imagine that, someday, my mortal remains, specks of dust, immersed in this heavenly wilderness, will be able to complete the circuit and loop around the Atlantic. I'll be in the good company of minute drifters, nomadic bluefin, and every adventurer who has traveled the Atlantic's river.

BIBLIOGRAPHY

General

Alexander, John, and James Lazell. *Ribbon of Sand: The Amazing Convergence of the Ocean and the Outer Banks*. Chapel Hill: University of North Carolina Press, 1992.

Boorstin, Daniel. *The Discovers*. New York: Random House, 1983.

Garrison, Tom. *Oceanography: An Invitation to Marine Science*. Pacific Grove: Wadsworth/Thomson Learning, 2002.

Karleskint, George, R. Turner, and J. Small. *Introduction to Marine Biology*. Belmont: Thomson Brooks/Cole, 2005.

Levinton, Jeffrey. *Marine Biology: Function, Biodiversity, Ecology*. New York: Oxford University Press, 2001.

MacLeish, William. *The Gulf Stream: Encounters with the Blue God*. New York: Houghton Mifflin, 1989.

———. "Painting a Portrait of the Stream from Miles Above—and Below." *Smithsonian*, February 1989, 42–55.

Mason, Paul. "The Changeable Ocean River." *Sea Frontiers*, March 1975, 171–77.

Morison, Samuel Eliot. *The Great Explorers: The European Discovery of America*. New York: Oxford University Press, 1978.

Nelson, Joseph. *Fishes of the World*. New York: John Wiley and Sons, 1994.

Pinet, Paul. *Oceanography: An Introduction to the Planet Oceanus*. St. Paul: West Publishing, 1992.

Richardson, Philip. "Tracking Ocean Eddies." *American Scientist*, May/June 1993, 261–71.

Schlee, Susan. *The Edge of an Unfamiliar World: A History of Oceanography*. New York: E. P. Dutton, 1973.

Stewart, R. W. "The Atmosphere and the Ocean." *Scientific American*, September 1969, 76–88.

Taylor, Kendrick. "Rapid Climate Change." *American Scientist*, July/August 1999, 320–25.

Webster, Peter, and Judith Curry. "The Oceans and Weather." In *Scientific American Presents: The Oceans*, 38–43. New York: Scientific American, 1998.

Wunsch, Carl. "What Is Thermohaline Circulation?" *Science* 229 (2002): 1179–81.

Chapter 1. Swirls and Conveyors

Andel, Tjeerd van. *Science at Sea: Tales on an Old Ocean*. San Francisco: W. H. Freeman, 1977.

Barth, Suzanna, and Bob Leben. "Navigating the Bermuda Race." <http://ccar.colorado.edu/~ altimetry/applications/sailing>. 17 September 2005.

Bourles, B., R. L. Molinari, E. Johns, W. D. Wilson, and K. D. Leaman. "On the Circulation in the Upper Layers of the Western Tropical North Atlantic." *Journal of Geophysical Research* 104 (1999): 21151–70.

Broecker, Wallace. "The Biggest Chill." *Natural History* 96 (1987): 74–82.

Ebbesmeyer, Curtis, and James Ingraham. "Pacific Toy Spill Fuels Ocean Current Pathways Research." *Eos*, 13 September 1994, 425–30.

Hogg, Nelson. "On the Transport of the Gulf Stream between Cape Hatteras and the Grand Banks." *Deep-Sea Research* 39 (1992): 123–46.

Huffman, E. E., and S. J. Worley. "An Investigation of the Circulation of the Gulf of Mexico." *Journal of Geophysical Research* 91 (1986): 14221–36.

Knauss, John. "A Note on the Transport of the Gulf Stream." *Deep Sea Research* 16 (1969): 117–23.

Maury, Matthew Fontaine. *The Physical Geography of the Sea and Its Meteorology*. New York: Dover, 2003.

McCarthy, M. S., and L. D. Talley. "The Subpolar Mode Water of the North Atlantic Ocean." *Journal of Physical Oceanography* 12 (1982): 1169–88.

Olson, Donald. "Rings in the Ocean." *Annual Review of Earth and Planetary Sciences* 19 (1991): 283–311.

Rahmstorf, Stefan. "Currents of Change: Investigating the Ocean's Role in Climate." <http://www.pik-potsdam.de/~ stefan/essay.html>. 13 November 2005.

Rossby, H. Thomas, and Peter Miller. "Ocean Eddies in the 1539 Carta Marina by Olaus Magnus." *Oceanography* 16 (2003): 77–88.

Savidge, Dana. "Gulf Stream Meander Propagation Past Cape Hatteras." *Journal of Physical Oceanography* 34 (2004): 2073–85.

Sedberry, George, ed. *Islands in the Stream: Oceanography and Fisheries of the Charleston Bump*. Bethesda: American Fisheries Society, 2001.

Stommel, Henry, and Arnold Arons. "On the Abyssal Circulation of the World's Oceans." Pt. I, "Stationary Planetary Flow Patterns on a Sphere." *Deep Sea Research* 6 (1960): 140–54.

Swallow, John, and Valentine Worthington. "Measurements of Deep Currents in the Western North Atlantic." *Nature* 179 (1957): 1183–84.

Thomson, Alastair, ed. *The Poems of Tennyson*. London: Routledge and Kegan Paul, 1986.

Wiebe, Peter. "Rings of the Gulf Stream." *Scientific American*, March 1982, 60–70.

Wust, G. *Stratification and Circulation in the Antillean Caribbean Basin*. Palisades: Columbia University Press, 1964.

Chapter 2. Anatomy of the Gulf Stream

Charnock, Henry. "Physics in Oceanography." *Physics Education* 15 (1980): 344–49.
Cione, J. J., S. Raman, and J. Pietrafesa. "The Effect of the Gulf Stream–Induced Baroclinicity on the U.S. East Coast Winter Cyclones." *Monthly Weather Review* 121 (1993): 1531–51.
Holt, T., and Sethu Raman. "Marine Boundary Layer Structure and Circulation in the Region of Offshore Re-development of a Cyclone during GALE." *Monthly Weather Review* 118 (1990): 392–410.
Pillsbury, John Elliott. *The Gulf Stream: Methods of Investigation and Results of Research*. Appendix no. 10: Report for 1890, U.S. Coast and Geodetic Survey. Washington: Government Printing Office, 1891.
Rediker, Marcus. *Between the Devil and the Deep Blue Sea*. Cambridge: Cambridge University Press, 1987.
Riehl, Herbert. *Climate and Weather in the Tropics*. New York: Academic Press, 1979.
Trenberth, Kevin, and Julie Caron. "Estimates of Meridional Atmosphere and Ocean Heat Transports." *Journal of Climate* 14 (2001): 3433–43.
Wood, Peter. *Weathering the Storm: Inside Winslow Homer's Gulf Stream*. Athens: University of Georgia Press, 2004.
Wunsch, Carl. "Observing Ocean Circulation from Space." *Oceanus* 35 (1992): 9–17.

Chapter 3. Flowing Down the Hill: The History of Ocean Circulation

Barnes, Jonathan, ed. *The Complete Works of Aristotle*. Princeton: Princeton University Press, 1995.
Burstyn, Harold. "The Deflecting Force and Coriolis." *Bulletin of the American Meteorological Society* 47 (1966): 890–91.
———. "Theories of Winds and Ocean Currents from the Discoveries to the End of the Seventeenth Century." *Terrae Incognitae* 3 (1971): 7–31.
Deacon, Margaret, ed. *A Treatise concerning the Motion of the Sea and Winds (1677), Isaak Vossius, together with Isaak Vossius, De Motu Marium et Ventorum (1663)*. Delmar: Scholars' Facsimile and Reprint, 1993.
Hogg, N. C., and R. X. Huarg, eds. *Collected Works of Henry Stommel*. Boston: American Meteorological Society, 1996.

Knauss, John. *Introduction to Physical Oceanography*. New York: Prentice Hall, 1978.

Peterson, R., L. Stramma, and G. Kortum. "Early Concepts and Charts of Ocean Circulation." *Progress in Oceanography* 37 (1996): 1–115.

Preston, Diane, and Michael Preston. *A Pirate of Exquisite Mind: The Life of William Dampier*. New York: Walker, 2004.

Richardson, Philip. "Benjamin Franklin and Timothy Folger's First Printed Chart of the Gulf Stream." *Science* 207 (1980): 643–45.

Shaw, Thomas, trans. *The Odyssey of Homer*. New York: Oxford University Press, 1932.

Sobel, Dava. *Longitude: The True Story of a Lone Genius Who Solved the Greatest Scientific Problem of His Time*. New York: Penguin Group, 1996.

Stommel, Henry. "The Gulf Stream: A Brief History of the Ideas concerning Its Cause." *Scientific Monthly* 70 (1950): 242–53.

———. "The Westward Intensification of Wind-Driven Currents." *Transactions of the American Geophysical Union* 29 (1948): 202–6.

Vorsey, Louis De. "Pioneer Charting of the Gulf Stream: The Contributions of Benjamin Franklin and William Gerard De Brahm." *Imago Mundi* 28 (1976): 105–20.

Chapter 4. Floaters and Drifters

Chisholm, S., P. Falkowski, and J. Cullen. "Dis-Crediting Ocean Fertilization." *Science* 294 (2001): 308–10.

Dooley, J. K. "Fishes Associated with the Pelagic Sargassum Complex, with a Discussion of the Sargassum Community." *Contributions in Marine Science* 16 (1972): 1–32.

Ferrell, David. "Sargassum: Weed of Life." *Marlin: The International Sport Fishing Magazine*, October/November 1999, 55–57.

Genthe, Henry. "The Sargasso Sea." *Smithsonian*, November 1998, 82–93.

Hardin, Garrett. "The Tragedy of the Commons." *Science* 162 (1968): 1243–48.

Jacobson, Jon, and Alison Rieser. "The Evolution of Ocean Law." In *Scientific American Presents: The Oceans*, 100–105. New York: Scientific American, 1998.

Johnston, Cheryl. "Small, Jellylike Creatures on Outer Banks Pose No Danger." *Virginian-Pilot*, 26 June 2004, 35.

Longhurst, Alan, and Daniel Pauly. *Ecology of Tropical Oceans*. New York: Academic Press, 1987.

National Coalition for Marine Conservation. "Offshore Habitat Gets Federal Protection." *NCMC Marine Bulletin*, Fall 2003, 5–6.

Schweid, Richard. *Consider the Eel*. Chapel Hill: University of North Carolina Press, 2002.

Scripps Institution of Oceanography. "Dinoflagellate Bioluminescence." <http://siobiolum.ucsd.edu/Dino_bl.html>. 2 March 2006.

Steinberg, Deborah, C. Carlson, N. Bates, S. Goldthwait, L. Madin, and A. Michaels. "Zooplankton Migration and the Active Transport of Dissolved Organic and Inorganic Carbon in the Sargasso Sea." *Deep Sea Research* 147 (2000): 137–58.

Swap, Robert, S. Ulanski, M. Cobbett, and M. Garstang. "Temporal and Spatial Characteristics of Saharan Dust Outbreaks." *Journal of Geophysical Research* 101 (1996): 4205–20.

Verne, Jules. *Twenty Thousand Leagues under the Sea*. New York: Barnes and Noble, 2005.

Chapter 5. Bluefin Tuna: The Great Migration

Block, Barbara, H. Dewar, C. Farwell, and E. Prince. "A New Satellite Technology for Tracking the Movements of Atlantic Bluefin Tuna." *Proceedings of the National Academy of Science* 95 (1998): 9384–89.

Buck, Eugene. "Atlantic Bluefin Tuna: International Management of a Shared Resource." *CRS Report 95-367* (1995): 1–23.

Carey, Frank. "Fishes with Warm Bodies." *Scientific American*, February 1973, 36–44.

Fogt, Jan. "A River Runs through It." *Sport Fishing*, May 2002, 32–36.

Gaffney, Rick. "Solving Pelagic Puzzles." *Sport Fishing*, February 2000, 80–88.

Hersey, John. *Blues*. New York: Vintage, 1987.

Joseph, James, W. Klawe, and P. Murphy. *Tuna and Billfish—Fish without a Country*. La Jolla: Inter-American Tropical Tuna Commission, 1988.

Koegler, John. "Highly Migratory Species Report." *Jersey Coast Anglers Association Newsletter*, April 1998, 1–3.

Lutcavage, Molly. "Bluefin Tuna Spawning in Central North Atlantic." *Pelagic Fisheries Research Program*, April–June 2001, 1–3.

Maggio, Theresa. *Mattanza Love and Death in the Sea of Sicily*. Cambridge, Mass.: Perseus, 2000.

Malmquist, David. "The Great Ocean Migration." *Currents*, Spring 1997, 43–50.

Murphy, Richard. "The Structure of the Pineal Organ of the Bluefin Tuna, *Thunnus Thynnus*." *Journal of Morphology* 133 (1971): 1–15.

National Research Council. *An Assessment of Atlantic Bluefin Tuna*. Washington: National Academy Press, 1994.

Partridge, Brian, J. Johansson, and J. Kalish. "The Structure of Schools of Giant Bluefin in Cape Cod Bay." *Environmental Biology Fishes* 9 (1983): 253–62.
Safina, Carl. *Song for the Blue Ocean*. New York: Henry Holt, 1997.
Sissenwine, Michael, P. Mace, J. Powers, and G. Scott. "A Commentary on Western Atlantic Bluefin Tuna Assessments." *Transactions of the American Fisheries Society* 127 (1998): 838–55.
Videler, John. *Fish Swimming*. London: Chapman and Hall, 1993.
Vogel, Steven. *Life in Moving Fluids: The Physical Biology of Flow*. Princeton: Princeton University Press, 1994.
Webb, Peter. "Form and Function in Fish Swimming." *Scientific American*, February 1984, 56–68.
Whynott, Douglas. *Giant Bluefin*. New York: Farrar, Straus and Giroux, 1995.
Worm, Boris, H. Lotze, and R. Myers. "Predator Diversity Hotspots in the Blue Ocean." *Archea* 100 (2003): 9884–88.

Chapter 6. Fishing the Blue Waters

Benetti, Daniel, R. W. Brill, and S. A. Kraul Jr. "The Standard Metabolic Rate of Dolphin Fish." *Journal of Fish Biology* 46 (1995): 987–96.
Benetti, Daniel, E. S. Iverson, and A. C. Ostrowski. "Growth Rates of Captive Dolphin, *Coryphaena Hippurus*, in Hawaii." *Fishery Bulletin* 93 (1995): 152–57.
Brill, Richard. "Selective Advantages Conferred by the High Performance Physiology of Tunas, Billfishes and Dolphin Fish." *Comparative Biochemistry and Physiology* 113A (1996): 3–15.
Brownlee, John. "Game Plan: Blackfin Blitz." *Sport Fishing*, April 2005, 44–48.
Carlson, Tom. *Hatteras Blues: A Story from the Edge of America*. Chapel Hill: University of North Carolina Press, 2005.
Cubbage, Jim. "The Story on Bananas and Bad Luck." <http://www.captjim.com/bananas.html>. 22 March 2006.
Fritsches, Kristin, R. Brill, and E. Warrant. "Warm Eyes Provide Superior Vision in Swordfishes." *Current Biology* 15 (2005): 55–58.
Fritsches, Kristin, and Eric Warrant. "Do Tuna and Billfish See Colors?" *Pelagic Fisheries Research Program Newsletter* 9 (2004): 1–4.
Gregory, Lane. "Cubans' Vessels Going Farther than They Do; The Empty Boats Have Been Seen Drifting off N.C. in the Gulf Stream." *Virginian-Pilot*, 14 September 1994, A5.
Grubb, Thomas. *The Mind of the Trout: A Cognitive Ecology for Biologists and Anglers*. Madison: University of Wisconsin Press, 2003.
Hahn, Andy. "Bird-Dogging: Fish Hounds Seek Help from Above." *Sport Fishing*, August 2005, 92–97.

Hemingway, Ernest. *The Old Man and the Sea*. New York: Simon and Schuster, 1995.

———. "On the Blue Water: A Gulf Stream Letter." *Esquire*, April 1936, 31.

Marshall, Justin. *Vision in Billfish*. Kona: Pacific Ocean Research Foundation, 1998.

Schaffner, Herbert. *Saltwater Game Fish of North America*. New York: Friedman/Fairfax Press, 1995.

Shettleworth, Sara. *Cognition, Evolution, and Behavior*. New York: Oxford University Press, 1998.

Wilson, R. P., P. Ryan, A. James, and M. Wilson. "Conspicuous Coloration May Enhance Capture in Some Piscivores." *Animal Behavior* 35 (1987): 1558–60.

Chapter 7. Exploration and Discovery

Abbott, Jacob. *Hernando Cortez*. Honolulu: University Press of the Pacific, 2002.

Armstrong, Wayne. "Floaters: Thousands of Tropical Seeds Take to the Ocean." *Sea Frontiers*, May/June 1994, 24–30.

Black, Clinton. *Pirates of the West Indies*. Cambridge: Cambridge University Press, 1989.

Bradford, Ernie. *A Wind from the North: The Life of Prince Henry the Navigator*. New York: Harcourt, Brace, 1960.

Campbell, Russ. "The Buccaneers." <http://www.it4biz.com/omnibus/PortOfCall/buccaneer.htm>. 11 October 2005.

DeBry, John. "Hurricane of 1715: Spanish Treasure Fleet Disaster, Discovery and Salvage." <http://www.hrd1715.com/1715_Story.html>. 11 October 2005.

Dunn, Richard, and Gary Nash. *Sugar and Slaves: The Rise of the Planter Class in the English West Indies, 1624–1713*. Chapel Hill: University of North Carolina Press, 2000.

Emmer, Peter. "The First Global War: The Dutch versus Iberia in Asia, Africa and the New World, 1590–1609." *Journal of Portuguese History*, Summer 2003, 1–14.

Exquemelin, Alexander. *Buccaneers of America*. Mineola, N.Y.: Dover, 2000.

Fernandez-Armesto, Felipe. *Pathfinders: A Global History of Exploration*. New York: W. W. Norton, 2006.

Fuson, Robert. *Juan Ponce de León and the Spanish Discovery of Puerto Rico and Florida*. Blacksburg, S.C.: McDonald and Woodward, 2000.

Hugh, Thomas. *Rivers of Gold: The Rise of the Spanish Empire from Columbus to Magellan*. New York: Random House, 2004.

Lane, Kris. *Pillaging the Empire: Piracy in the Americas, 1500–1750*. Armonk: M. E. Sharpe, 1998.
Marden, Luis. "The First Landfall of Columbus." *National Geographic*, November 1985, 572–77.
Marx, Robert. *The Treasure Fleets of the Spanish Main*. New York: World, 1968.
Molander, Arne, and James Norris. "Did Seaweed Lead Columbus to the Antilles?" *Encounters*, nos. 5–6 (1990): 49.
Morison, Samuel. *The European Discovery of America: The Northern Voyages*. New York: Oxford University Press, 1971.
Mulcahy, Matthew. *Hurricanes and Society in the British Greater Caribbean, 1624–1783*. Baltimore: Johns Hopkins University Press, 2005.
Peck, Douglas. "Re-thinking the Columbus Landfall Problem." *Terrae Incognitae* 28 (1996): 12–35.
Prestage, Edgar. *The Portuguese Pioneers*. London: Adam and Charles Black, 1933.
Rawley, James. *The Transatlantic Slave Trade*. New York: W. W. Norton, 1981.
Richardson, Philip, and Roger Goldsmith. "The Columbus Landfall: Voyage Track Corrected for Winds and Currents." *Oceanus* 30 (1987): 3–10.
Rogozinski, Jan. *Brief History of the Caribbean: From the Arawak and Carib to the Present*. New York: Penguin Group, 1999.
Ruddiman, William. *Earth's Climate: Past and Future*. New York: W. H. Freeman, 2000.
Walton, Timothy. *The Spanish Treasure Fleets*. Sarasota: Pineapple Press, 1994.
Wilkinson, Jerry. "History of Wrecking." <http://www.keyshistory.org/wrecking.html>. 14 October 2005.
Williams, Eric. *From Columbus to Castro: The History of the Caribbean, 1492–1969*. New York: Harper and Row, 1971.
Williamson, J. A. *The Cabot Voyages and Bristol Discovery under Henry VII, with the Cartography of Voyages by R. A. Skelton*. New York: Cambridge University Press, 1962.

Chapter 8. Colonization of America

Alderman, Richard. *Rum, Slaves and Molasses: The Story of New England's Triangular Trade*. New York: Crowell-Collier Press, 1972.
Ashley, Clifford. *The Yankee Whaler*. New York: Dover, 1926.
Barnes, Jay. *North Carolina's Hurricane History*. Chapel Hill: University of North Carolina Press, 1995.
Black, Peter, and Lawrence Shay. "Observed Sea Surface Temperature Variability in Tropical Cyclones: Implications for Structure and Intensity Change."

Preprint, 21st Conference on Hurricane and Tropical Meteorology, 24–28 April 1995, Miami, Fla., 603–4.

Butler, Lindley. *Pirates, Privateers, and Rebel Raiders of the North Carolina Coast*. Chapel Hill: University of North Carolina Press, 2000.

Caffrey, Kate. *The Mayflower*. New York: Stein and Day, 1974.

Chadwick, Ian. "Henry Hudson's Third Voyage, 1609: The New World." <http://www.ianchadwick.com/hudson/hudson_03.htm>. 15 December 2006.

Chamberlain, Barbara. *These Fragile Outposts: A Geological Look at Cape Cod, Martha's Vineyard, and Nantucket*. Garden City: Natural History Press, 1964.

Chatterton, E. Keble. *English Seamen and the Colonization of America*. London: Arrowsmith, 1930.

Creighton, Margaret. *Rites and Passages: The Experience of American Whaling, 1830–1870*. New York: Cambridge University Press, 1995.

Curtin, Philip. *The Rise and Fall of Plantation Complexes: Essays in Atlantic History*. Cambridge: Cambridge University Press, 1990.

Ellis, Richard. *Men and Whales*. New York: Knopf, 1991.

Gareth, Rees. "Copper Sheathing: An Example of Technological Diffusion in the English Merchant Fleet." *Journal of Transport History*, n.s., 1 (1971): 85–94.

Howe, Octavius, and Fredrick Matthew. *American Clipper Ships, 1833–1858*. Mineola, N.Y.: Dover, 1986.

Israel, Jonathan. *Dutch Primacy in World Trade, 1585–1740*. London: Oxford University Press, 1990.

Johnson, Captain Charles. [Daniel Defoe]. *A General History of The Pyrates*. Mineola, N.Y.: Dover, 1999.

Johnson, Donald. *Charting the Sea of Darkness: The Four Voyages of Henry Hudson*. New York: McGraw-Hill, 1992.

Kurlansky, Mark. *Cod: A Biography of the Fish That Changed the World*. New York: Penguin Books, 1997.

Ludlum, David. *Early American Hurricanes, 1492–1870*. Boston: American Meteorological Society, 1963.

Mountford, Kent. "History behind Sugar Trade, Chesapeake Not Always Sweet." *Bay Journal*, July/August 2003, 13–15.

Philbrick, Nathan. *Mayflower: A Story of Courage, Community, and War*. New York: Viking, 2006.

Price, David. *Love and Hate in Jamestown: John Smith, Pocahontas, and the Heart of the New World*. New York: Knopf, 2003.

Roosevelt, Theodore. *New York: A Sketch of the City's Social, Political, and Commercial Progress from the First Dutch Settlements to Recent Times*. New York: Charles Scribner and Sons, 1906.

Santos, Robert. "Azorean and New England Whaling and Fishing." <http://library.csustan.edu/bsantos/whaling.html>. 9 October 2006.

Sayles, T. F. "Just Imagine." *Chesapeake Bay Magazine*, August 2006, 8.

Schoenbaun, Thomas. *Islands, Capes, and Sounds: The North Carolina Coast.* Winston-Salem: John F. Blair, 1982.

Sirmans, Eugene. *Colonial South Carolina: A Political History, 1663–1763.* Chapel Hill: University of North Carolina Press, 1966.

Tickner, F., and V. Medvei. "Scurvy and the Health of European Crews in the Indian Ocean in the Seventeenth Century." *Medical History* 2 (1958): 36–46.

Waring, Gordon, C. Fairfield, C. Ruhsam, and M. Sano. "Sperm Whales Associated with Gulf Stream Features off the Northeastern USA Shelf." *Fisheries Oceanography* 2 (1993): 101–5.

INDEX